Combinatorics
A Problem Oriented Approach

© 1998 by

The Mathematical Association of America (Incorporated)

Library of Congress Catalog Card Number 98-85594

ISBN 978-0-88385-710-6

Printed in the United States of America

Current Printing (last digit):
10 9 8 7 6 5 4 3 2

Combinatorics
A Problem Oriented Approach

Daniel A. Marcus
California State Polytechnic University, Pomona

Published and Distributed by
THE MATHEMATICAL ASSOCIATION OF AMERICA

Classroom Resource Materials is intended to provide supplementary classroom material for students—laboratory exercises, projects, historical information, textbooks with unusual approaches for presenting mathematical ideas, career information, etc.

Committee on Publications
James W. Daniel, *Chair*

Andrew Sterrett, Jr., *Editor*

Frank Farris	Edward M. Harris
Yvette C. Hester	Millianne Lehmann
Dana N. Mackenzie	Edward P. Merkes
William A. Marion	Daniel Otero
Alec Norton	Barbara J. Pence
Dorothy D. Sherling	Michael Starbird

101 Careers in Mathematics, edited by Andrew Sterrett
Calculus Mysteries and Thrillers, R. Grant Woods
Combinatorics: A Problem Oriented Approach, Daniel A. Marcus
Elementary Mathematical Models, Dan Kalman
Interdisciplinary Lively Application Projects, edited by Chris Arney
Laboratory Experiences in Group Theory, Ellen Maycock Parker
Learn from the Masters, Frank Swetz, John Fauvel, Otto Bekken,
      Bengt Johansson, and Victor Katz
Mathematical Modeling for the Environment, Charles Hadlock
A Primer of Abstract Mathematics, Robert B. Ash
Proofs Without Words, Roger B. Nelsen
A Radical Approach to Real Analysis, David M. Bressoud
She Does Math!, edited by Marla Parker

MAA Service Center
P. O. Box 91112
Washington, DC 20090-1112
1-800-331-1622   fax: 1-301-206-9789

For Shelley, Alina, Larry, and Andy

Contents

Preface

This book developed from a course in combinatorics that I have taught for more than twelve years at California State Polytechnic University. My classes are populated primarily by mathematics and computer science majors, with a smaller number of engineering and other science students. The class level is generally third and fourth year, but there are no particular prerequisites. I tell students just to bring their brains.

The format is unique, in that it combines features of a traditional textbook with those of a problem book. The material is presented through a series of approximately 250 problems with connecting text, supplemented by approximately 220 additional problems for homework assignments. The problems are arranged strategically to introduce concepts in a logical order and in a provocative way. My lectures usually consist of working problems at the board with class input, but there are other possible approaches: students might work problems in groups, or be assigned a series of problems to present, in effect delivering part of the lecture.

The book is divided into eight sections, the first four of which cover the basic combinatorial entities of strings, combinations, distributions, and partitions. The last four cover the special counting methods of inclusion and exclusion, recurrence relations, generating functions, and the method of Polya and Redfield that can be characterized as "counting modulo symmetry." Each section is divided into two groups of problems, roughly identifiable as classroom problems and homework problems, with the latter group beginning at the heading **More Problems**. By using the "Dependence of Problems" table on page 123, one can determine how far into a particular section the lecture must progress before a given homework problem can be assigned. At the end of the book are answers to selected classroom problems; these problems have been identified in the text by an asterisk.

PART I

Basics

A

Strings

A *string* is a list, or sequence, of elements in a particular order. For the time being, we will concern ourselves with finite strings only.

Examples $(1, 3, 0, 1, -12)$ is a string of integers. (X, Q, R, Z, X) is a string of letters from the usual alphabet $\{A, B, C, \ldots, Z\}$

Notice that the same element can appear more than once in a string.

When no confusion results, we will usually leave out the parentheses and commas in the representation of a string. The latter example above can be written more easily as *XQRZX*. This five-letter string can also be thought of as a "word" that uses letters from the alphabet $\{A, \ldots, Z\}$. The terms *string*, *sequence*, and *word* will be used interchangeably.

Example Any positive integer can be represented by a string of digits from the set $\{0, 1, 2, \ldots, 9\}$. This is the usual base 10 representation. In the binary system, any positive integer is represented by a string of digits (called *bits* in this case) from the set $\{0, 1\}$.

Counting Strings

Continuing with the example above, consider all of the integers from 0 to 999. Each one of these can be represented by a 3-digit string using digits from 0

3

to 9; 0 is represented as 000, 1 as 001, etc. We know that there are 1000 of these integers (0 to 999), and the same result can be obtained by counting 3-digit strings: The first digit can be selected in 10 different ways, and for each of these choices there are 10 ways to select the second digit. This gives $10 \cdot 10$, or 100, ways to select the first two digits. Finally, each of these 100 choices can be combined with one of the 10 choices for the third digit. The result is 1000 different three-digit strings.

Here are some problems for you to try.

A1 By counting choices as in the example above, determine the number of different five-bit strings, which are strings of 0s and 1s, each consisting of five bits. What does this have to do with the binary representation of integers?

A2 Find the number of five-letter words using letters of the alphabet $\{A, B, C, \ldots, Z\}$.

These are instances of a more general problem, which can be summarized as follows.

Standard Problem #1
Find the number of strings of a given length that use elements from a given set.

The *length* of a string means the number of terms, or elements, in the string. The answer to Standard Problem #1 is n^k, where k is the length of the string and n is the number of elements in the set from which the terms are selected.

Next, we look at some variations on this standard problem. Suppose we want to count three-letter words that use letters from the usual alphabet, but with the condition that no letter can occur more than once in each word. Again, we approach the problem by counting choices for each letter: As in problem A2, there are 26 choices for the first letter. For each such choice there are only 25 letters that can be placed in the second position, because the letter used in the first position cannot be repeated. This gives $26 \cdot 25$ ways to place two different letters in the first two positions, and each of these choices can be combined with one of the remaining 24 letters in the third position. The resulting number of three-letter words is $26 \cdot 25 \cdot 24$.

A3 Find the number of five-digit strings using digits from $\{0, 1, 2, \ldots, 9\}$ if no digit appears more than once in the string.

Standard Problem #2
Find the number of strings of a given length that use elements from a given set, if no element appears more than once in any string.

In other words, we are counting strings with no repeated elements. If k is the length of the string and n is the number of elements in the given set, then the answer is:

$$n(n-1)(n-2)\cdots(n-k+1).$$

(We assume that $n \geq k$.) An alternative form of this answer is

$$\frac{n!}{(n-k)!}$$

where, in the case $n = k$, we recall that $0! = 1$.

Permutations

A *permutation* of a set is an arrangement of all of the elements of the set in a particular order. In other words, it is a string in which each element of the set appears exactly once.

A4   Find the number of permutations of each set below.

(a) $\{A, B, C\}$

(b) $\{0, 1, 2, \ldots, 9\}$

(c) $\{A, B, C, \ldots, Z\}$

Counting permutations is a special case of Standard Problem #2, in which $k = n$. The result is $n!$ for a set with n elements.

A5   How many ways can the letters of the word *FNORG* be arranged?

A6*   How many ways can the letters of the word *XQRZX* be arranged? (*Hint:* First label the two Xs as X_1 and X_2 and consider them different. What is the result? More on this type of problem will appear in Section C.)

Not all problems fit neatly into one of these standard categories. Often, it is necessary to use the more general formula that follows.

Product Rule   In a string of length k, if the ith term can be filled in a particular number of ways (n_i), which does not depend on how any of the previous terms have been filled, then the total number of possible strings is the product $n_1 n_2 n_3 \cdots n_k$.

A7   Find the number of sequences of length 4 in which the first two terms are letters from the alphabet $\{A, B, C, \ldots, Z\}$ and the last two terms are digits from the set $\{0, 1, 2, 3, \ldots, 9\}$, with the condition that the two digits must be different.

A8* Find the number of five-letter words using letters from $\{A, B, C, \ldots, Z\}$ in which no two consecutive letters are the same.

Further complications can occur. The problems that follow must be viewed in a slightly different way.

A9 Find the number of five-letter words that use letters from $\{A, B, C, \ldots, Z\}$ in which the letter A occurs exactly once. (*Hint:* Where does A occur? Consider the possibilities separately.)

A10 Find the number of five-letter words that use letters from $\{A, B, C, \ldots, Z\}$ in which the letter A occurs *at least* once. (*Hint:* First count words that don't contain any As.)

A11* Find the number of five-letter words that use letters from the three-letter set $\{A, B, C\}$ in which each letter occurs at least once. (*Hint:* As in the hint for problem A10, first count the words that you *don't* want to include.)

A12 Find the number of four-letter words that use letters from $\{A, B, C\}$ in which no three consecutive letters are all the same. (Same hint as above.)

A13* Find the number of three-letter words that use letters from the ten-letter set $\{A, B, C, \ldots, J\}$ in which all letters are different and the letters appear in alphabetical order. (*Hint:* First count the three-letter words in which all of the letters are different. Then explain why exactly one-sixth of these words have their letters in alphabetical order.)

A14 Find the number of strictly increasing three-digit strings that use digits from $\{0, 1, 2, \ldots, 9\}$. (In other words, count strings abc in which $a < b < c$. *Hint:* Does this have something to do with the preceding problem?)

Probability

Suppose that we select a five-letter word at random from the set of all five-letter words that use letters from $\{A, B, C\}$. What is the probability that the word selected does not contain the letter A? Assuming that all 3^5 possible words are equally likely to be selected, the answer depends on how many of these words do not contain A. The number of such words is 2^5, so it follows that the probability is the ratio $2^5/3^5$. In general, if there are n equally likely ways in which something can happen and exactly m of these satisfy a certain condition, then the probability of this condition being satisfied is m/n.

A15 Find the probability that a five-letter word that uses letters from the set $\{A, B, C\}$ contains at least one of each letter. (*Hint:* Use your answer from problem A11.)

A16 Find the probability that a four-letter word that uses letters from $\{A, B, C, D, E\}$ contains

(a) no repeated letters;

(b) no two consecutive equal letters.

A17 Suppose the numbers $1, 2, 3, 4, 5$ are arranged in a random order. Find the probability that all odd numbers still appear in odd-numbered positions.

A18* Four cards are selected at random from a standard deck of 52 cards. Find the probability that

(a) all of them are aces;

(b) they are from four different suits.

(*Suggestion:* Select the cards one at a time and count sequences of cards.)

A19 (a) A coin is flipped four times. Find probability that it lands "heads" twice. (*Hint:* Count strings of length 4 consisting of the letters H and T. How many of these strings contain exactly two Hs?)

(b) Four coins are flipped simultaneously. Find the probability that exactly two of them land "heads."

Rearrangements and Derangements

Definitions A *rearrangement* of a string is any ordering of the elements of the string, including the original ordering. A *derangement* of a string of distinct elements is a rearrangement of the string such that no element appears in its original position.

Examples *BCA* is a derangement of *ABC*.
ABC is a derangement of *BCA*.
53214 is a derangement of 12345.

A20   Find all of the derangements of

(a) *AB*

(b) *ABC*

(c) *ABCD*

(d) 321

Notation D_n denotes the number of derangements of a string of n distinct elements.

From problem A20, we know $D_2 = 1, D_3 = 2, D_4 = 9$. In sections E and F, and we will find out how to calculate more of these numbers. The next one is $D_5 = 44$.

A21   Find the number of rearrangements of *ABCDE* having exactly one letter in its original position. (*Hint:* First suppose it is *A*.)

A22*   Find the number of four-letter words that use letters from the set $\{A, B, C, D\}$ such that the first letter is not *A*, the second is not *B*, the third is not *C*, and the fourth is not *D*.

A23*   Find the number of rearrangements of 1234 such that 1 is not in position 3, 2 is not in position 1, 3 is not in position 2, and 4 is not in position 4.

Section Summary

A *string* is a list, or sequence, of elements in a particular order. The *length* of a string is the number of terms. There are n^k strings of length k using elements from an n-element set. The number of such strings in which no element occurs more than once is $n(n - 1) \cdots (n - k + 1)$, or $n!/(n - k)!$. More generally, the *product rule* counts strings of length k in which the ith term can be filled in n_i ways: The number of such strings is $n_1 n_2 \cdots n_k$.

The *probability* of an event is a fraction defined as follows: If there are n equally likely ways in which something can happen and exactly m of these satisfy a certain condition, then the probability of this condition being satisfied is m/n.

A *rearrangement* of a string is any ordering of the elements of the string, including the original ordering. A *derangement* is a rearrangement of a string of distinct elements in which no element occupies its original position. The number D_n counts the derangements of a string of length n. The first few derangement numbers are $D_2 = 1, D_3 = 2, D_4 = 9$, and $D_5 = 44$.

More Problems

A24 Find the number of possible combinations for a combination lock if each combination consists of three integers, not necessarily distinct, from 0 to 39. (*Warning:* The term "combination" will take on a special mathematical meaning in the next section. Do not confuse the ordinary use in this problem with that meaning.)

A25 Find the number of seven-letter words that use letters from the alphabet $\{A, B, C, \ldots, Z\}$ in which no letter is repeated.

A26 Find the number of possible radio station names, if the name must contain either three or four letters and the first letter must be either K or W.

A27 (a) Find the number of five-digit strings using digits from $\{0, 1, 2, \ldots, 9\}$ that contain at least one 7. (*Hint:* How many do not?)
(b) Find the number of positive integers $<100{,}000$ that contain at least one 7.

A28 Find the number of three-digit numbers (integers from 100 to 999) that contain no two consecutive equal digits.

A29 (a) Find the number of bit strings (strings using 0s and 1s) of length 26.
(b) Find the number of subsets of the alphabet $\{A, B, C, \ldots, Z\}$. What is the connection with part (a)?

A30 (a) Find the number of ways to seat seven people around a circular table if all rotations of a particular arrangement are considered to be the same.
(b) How many ways can seven keys be put on a circular key ring? (*Note:* The essential difference between keys on a ring and people around a table is that the keys will not object if the entire ring is turned upside down.)
(c) Suppose the key ring in (b) has a chain attached to it somewhere. How does that change the answer?

A31 Find the number of five-letter words that use letters from the alphabet $\{A, B, C, \ldots, Z\}$ in which every sequence of three consecutive letters includes three different letters.

A32 Find the number of five-letter words that use letters from the set $\{A, B, C\}$ and include at least one A and at least one B.

A33 (a) Find the number of ways to place eight identical objects on an eight-by-eight chessboard so that no two of them are in the same row or column. (*Suggestion:* If necessary, look first at a simpler case, such as a four-by-four board.)

(b) Suppose that the objects in (a) are all considered to be different. Then what is the result?

A34 Find the number of permutations of the digits $0, 1, \ldots, 9$ in which odd and even digits alternate.

A35 Find the number of rearrangements of the word *ABCDEFG* that contain each of the following.

(a) The sequence *ABC*. (*Hint:* Glue some letters together.)

(b) The sequences *AB*, *CD*, and *EF*.

(c) The sequences *AB*, *BC*, and *EF*.

A36 Find the number of five-letter words that use letters from $\{A, B, C, \ldots, Z\}$ in which all letters are different and are in alphabetical order.

A37 Find the probability that a five-letter word that uses letters from $\{A, B, C\}$ contains exactly one *A*.

A38 If five coins are flipped simultaneously, find the probability of each of the following:

(a) At least one coin lands "heads;"

(b) At most one coin lands "heads."

A39 Seven people are seated in a row. They all get up and sit down again in random order. What is the probability that the two people originally seated at the two ends are no longer at the ends after they sit down again?

A40 Three cards are selected at random from a standard deck of 52 cards. Find the probability of each of the following:

(a) All of them are from the same suit.

(b) All of them are from different suits.

A41 From a set of five pairs of shoes, two of the shoes are selected at random. Find the probability of each of the following:

(a) Both are from the same pair.

(b) One left shoe and one right shoe are selected.

A42 Suppose that ten days of the year, not including February 29, are chosen at random in a particular order, forming a sequence. Repeated days are allowed.

(a) Count the sequences in which all ten days are different.

(b) Find the probability that all ten days are different.

A43   Assume that a group of people contains no one whose birthday is February 29.

(a) If there are ten people in the group, find the probability that at least two of them have the same birthday (month and day).

(b) How large must a group be so that there is a greater than 50% chance that at least two members have the same birthday?

A44   If a positive integer n is factored into primes in the form $p_1^{a_1} p_2^{a_2} \ldots$, then the divisors of n have the form $p_1^{b_1} p_2^{b_2} \ldots$, where each exponent b_i can range from 0 to a_i. For example, the divisors of 12 (which is $2^2 3^1$) have the form $2^{b_1} 3^{b_2}$, where $b_1 = 0$, 1 or 2, and $b_2 = 0$ or 1. This fact allows us to count the number of divisors of a given integer.

(a) List the six divisors of 12, including 1 and 12, and explain combinatorially (that is, by a counting procedure established in this section) why there are six of them.

(b) Without actually listing them, find the number of divisors of each of the integers 77, 100, 800, and 3000.

A45   (a) If a three-letter word that uses letters from $\{A, B, C\}$ is selected at random, find the probability that the first letter is not A, the second is not B, and the third is not C.

(b) If a rearrangement of ABC is selected at random, find the probability that the first letter is not A, the second is not B, and the third is not C.

A46   Decide what we would mean by a derangement of a string in which the elements are not all distinct. Find the number of derangements of each of the following strings.

(a) *AAB*

(b) *AABB*

(c) *AABC*

(d) *AABBC*

A47   Find the number of derangements of each of the following strings:

(a) *AABCD*

(b) *AABBCC*

B

Combinations

The term *combination* applies to a list of elements when the order of the elements is not taken into account.

Example   The two words *ABC* and *BCA* both represent the same combination of letters.

Elements in a combination can be distinct, as in the preceding example, or can repeat.

Example   *AABC* is the same combination as *ABCA*, but is different from *ABCC*.

B1   Find the number of two-letter combinations that use letters from the set $\{A, B, C, D, E\}$, with each of the following conditions.

(a) No letters can be repeated.

(b) Letters can be repeated.

B2*   Find the number of three-letter combinations that use letters from the set $\{A, B, C, D, E\}$, with each of the following conditions.

(a) No letters can be repeated. (*Hint:* Find a connection with (a) of problem B1.)

(b) Letters can be repeated.

13

B3 Find the number of three-digit combinations from the set $\{0, 1, 2, \ldots, 9\}$ in which no digit is repeated. Do this by the following sequence of steps.

(a) Count the number of three-digit strings that contain no repeated digits.

(b) Explain why each combination that we want to count corresponds to exactly six of these strings.

(c) Use (a) and (b) to determine the number of combinations.

Standard Problem #3

Find the number of combinations of a given length that consist of distinct elements from a given set.

By the *length* of a combination, we mean the number of elements, including any repetitions. To solve this problem, let n be the number of elements in the given set and let k be the number of elements in the combination. First, count the strings of length k from this set that have no repeated elements. This was Standard Problem #2, and the number of such strings is

$$n(n - 1)(n - 2) \cdots (n - k + 1).$$

Each combination that we want to count occurs a number of times among these strings. Specifically, each combination can be rearranged in $k!$ ways. Each of these rearrangements represents a different string but the same combination, so the product above counts each combination exactly $k!$ times. Therefore the answer to Standard Problem #3 is:

$$\frac{n(n - 1)(n - 2) \cdots (n - k + 1)}{k!}$$

where k is the length of the combination and n is number of elements in the set; or, in another form,

$$\frac{n!}{k!\,(n - k)!}$$

This number is written as $C(n, k)$ or $\binom{n}{k}$ and is referred to as "n choose k" since it counts the number of ways to choose k distinct elements from a set of n elements, assuming that the order in which the elements are chosen is not taken into account. These numbers are called *combination numbers*.

B4 (a) Find the number of ten-element combinations without repetition from a 100-element set.

(b) Find the number of 90-element combinations without repetition from a 100-element set.

(c) In general, what is the relationship between the numbers $\binom{n}{k}$ and $\binom{n}{n-k}$?

B5 Solve problem A18 again, this time by counting combinations of cards.

Pascal's Triangle

The combination numbers $\binom{n}{k}$ are the same numbers that appear on the well-known "Pascal's triangle,"

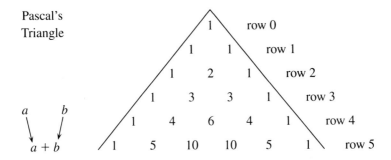

in which each element is the sum of the two numbers diagonally above it. By starting the numbering of the rows with 0, we can say that $\binom{n}{k}$ appears in row n. Row n consists of the combination numbers

$$\binom{n}{0}\binom{n}{1}\binom{n}{2}\cdots\binom{n}{n-1}\binom{n}{n}.$$

If we split the triangle into diagonals as shown in the following figure, then $\binom{n}{k}$ appears in the kth diagonal, where the numbering again begins at 0.

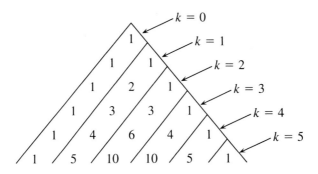

B6*   Extend Pascal's triangle to include row 9.

B7   (a) What seems to be true about the sum of all of the numbers in row n of Pascal's triangle?

(b) Explain why this happens, in terms of the way the triangle is formed.

(c) Explain why this happens, in terms of the fact that the combination numbers count subsets of a set.

We have not yet explained why the combination numbers appear on Pascal's triangle. At the extremes, it is easy to see why $\binom{n}{0} = \binom{n}{n} = 1$ and $\binom{n}{1} = \binom{n}{n-1} = n$. But beyond that, it is not so obvious; what is missing is an explanation of why the combination numbers can be generated in the same way that the triangle is generated. Specifically, we want to show that

$$\binom{n}{k} + \binom{n}{k+1} = \binom{n+1}{k+1}$$

This can be done in two different ways. The first is algebraic, and is illustrated by the following problem.

B8   Complete the following calculation, showing that the sum is equal to $\binom{n+1}{k+1}$.

$$\frac{n!}{k!\,(n-k)!} + \frac{n!}{(k+1)!\,(n-k-1)!} =$$

The second method is combinatorial. Consider the set of $n+1$ integers $\{0, 1, 2, \ldots, n\}$ and count the combinations consisting of $k+1$ distinct elements from this set. The number of such combinations is $\binom{n+1}{k+1}$.

B9   Among these subsets, some include 0 and others do not. How many are there of each type?

Sometimes combination numbers turn up in situations where we might not expect them.

B10*   Find the number of seven-letter words that use letters from the two-letter set $\{A, B\}$ if each word must include exactly three As. (*Hint:* This has something to do with subsets of the set $\{1, 2, 3, 4, 5, 6, 7\}$.)

B11   Find the number of rearrangements of the bit string 0011000101. (*Reminder:* We use the term *rearrangement* to refer to any ordering of the elements of a word or string, including the original ordering.)

Binomial Expansions

If we expand the expression $(A + B)^4$ by first writing

$$(A + B)(A + B)(A + B)(A + B),$$

we find that each term in the product comes from selecting either A or B from each of the four factors and multiplying them. The result is:

$$AAAA + AAAB + AABA + ABAA + BAAA + AABB + \cdots + BBBB$$
$$= A^4 + 4A^3B + 6A^2B^2 + 4AB^3 + B^4.$$

Notice that the coefficient of A^3B represents the number of rearrangements of three As and one B. In the next term, the coefficient of A^2B^2 is the number of rearrangements of two As and two Bs. This is $\binom{4}{2} = 6$, since the two As can occupy any two of the four positions.

In general, the coefficient of A^kB^{n-k} in the expansion of $(A + B)^n$ is $\binom{n}{k}$. So we can write

$$(A + B)^n = \sum_{k=0}^{n} \binom{n}{k} A^k B^{n-k}$$

This is the *binomial expansion formula*.

B12 Find the coefficient of X^7Y^{13} in the expansion of each of the following.

(a) $(X + Y)^{20}$

(b) $(3X–5Y)^{20}$

B13* Find the coefficient of X^8 in the expansion of each of the following.

(a) $(X–1)^{15}$

(b) $(X^2 + 1)^{15}$

B14 Use the binomial expansion formula to find the alternating sum of the numbers in row n of Pascal's triangle:

$$\binom{n}{0} - \binom{n}{1} + \binom{n}{2} - \cdots$$

(*Hint:* Select A and B appropriately.)

We return to counting rearrangements, but with an added restriction.

B15 Find the number of ways to rearrange three As and seven Bs if no two As can appear consecutively. (*Hint:* Start with *BBBBBBB* and consider where the three As can be inserted. There are eight places.)

B16* Use the method suggested by problem B15 to fill in the answer to the following standard problem.

Standard Problem #4
Find the number of bit strings that contain a given number of 0s and 1s, such that there are no two consecutive 1s.

Let m be the number of 0s and n be the number of 1s. What is the answer, in terms of m and n?

B17* Find the number of ways to select three distinct digits from the set $\{0, 1, 2, \ldots, 9\}$ if no two consecutive digits can be selected. (*Hint:* This has something to do with problem B15.)

Combinations Allowing Repeptition

Next, we look at combinations from a given set in which repeated elements are allowed to occur. We begin by assuming that each element in the given set occurs at least once in the combination, and we see how to count combinations of this type. Then we will be able to handle the more general case.

Consider the following problem. Find all five-digit combinations from the set $\{1, 2, 3\}$ with no missing digits. This means that each digit occurs at least once in each combination. Counting combinations of this type by listing them, we find six.

$$11123$$
$$11223$$
$$11233$$
$$12223$$
$$12233$$
$$12333$$

Notice what happens when the symbol "$<$" is inserted in the appropriate places in each combination.

1	1	1	< 2	< 3
1	1	< 2	2	< 3
1	1	< 2	< 3	3
1	< 2	2	2	< 3
1	< 2	2	< 3	3
1	< 2	< 3	3	3

The number of combinations is $\binom{4}{2} = 6$, corresponding to the number of ways to select two out of four positions for the $<$ signs. Any choice of two positions determines one five-digit combination from $\{1, 2, 3\}$ with no missing digits. For example, in the case below, notice that there is only one way to fill in the digits $1, 2, 3$ in the five boxes. (We assume that the digits appear in nondecreasing order. Each combination has exactly one representation in that form.)

B18   Find the number of 50-digit combinations from $\{0, 1, 2, \ldots, 9\}$ with no missing digits.

We can now solve the following standard problem.

Standard Problem #5
Find the number of combinations of a given length that use elements from a given set, allowing repetition and with no missing elements.

(In other words, each element of the given set occurs at least once in each combination.)

The answer is $\binom{k-1}{n-1}$, where n is the number of elements in the given set and k is the length of the combination. Each combination corresponds to a choice of $n - 1$ locations for $<$ signs in the $k - 1$ spaces between elements in the combination. We assumed for the purpose of this explanation that the given set consists of the integers $\{1, 2, 3, \ldots, n\}$, but the same result applies to any set of n elements.

B19   Find the number of 100-letter combinations from the alphabet $\{A, B, C, \ldots, Z\}$ with no missing letters.

B20*   Find the number of 50-digit combinations from $\{0, 1, 2, \ldots, 9\}$ in which each digit occurs at least two times. (*Hint:* Start by including one of each digit and consider how many ways the remaining elements of the combination can be selected.)

B21   Find the number of 20-letter combinations from the set $\{A, B, C\}$ containing at least one A, at least two Bs, and at least three Cs.

Suppose we want to count combinations allowing repetition and also allowing missing elements. For example, consider the two-letter combinations from the set $\{A, B, C\}$.

$$AA, \ BB, \ CC, \ AB, \ AC, \ BC$$

If we add one instance of each letter to each of these six combinations, we get all of the five-letter combinations of A, B, C with no missing letters.

$$AA \ ABC = AAABC$$

$$BB \ ABC = ABBBC$$

$$CC \ ABC = ABCCC$$

$$AB \ ABC = AABBC$$

$$AC \ ABC = AABCC$$

$$BC \ ABC = ABBCC$$

This example shows how to convert the problem of counting combinations allowing missing letters into the form of Standard Problem #5. We know there are $\binom{4}{2} = 6$ five-letter combinations with no missing letters, so there must be an equal number of two-letter combinations allowing missing letters.

Try the following similar problem.

B22   Find the number of three-letter combinations from the set $\{A, B, C, D, E\}$ allowing repetition.

(*Note:* In general, if missing elements are not explicitly prohibited we will assume they are allowed.)

Standard Problem #6

Find the number of combinations of a given length that use elements from a given set, allowing repetition.

The answer is $\binom{n+k-1}{n-1}$ or, equivalently, $\binom{n+k-1}{k}$ where n is number of elements in the given set and k is the length of the combination (notice why the two expressions are equal).

This is explained as follows: If we add one of each element to each combination, then we obtain combinations of $n + k$ elements from the same set, allowing repetition and with no missing elements. The number of these is given by Standard Problem #5.

Notation   Combinations allowing repetition:

$$\binom{n}{k}_R = \binom{n + k - 1}{k}$$

B23   Find the number of five-letter combinations from the alphabet $\{A, B, C,$ $\ldots, Z\}$ allowing repetition.

B24*   Find the number of nondecreasing sequences of length 10

$$a_1 \leq a_2 \leq a_3 \leq \cdots \leq a_{10}$$

consisting of integers from the set $\{1, 2, 3, \ldots, 100\}$.

Consistently Dominated Sequences

The final topic in this section concerns sequences of elements from a given set. Such a sequence is *consistently dominated* if there is an element A such that within every initial segment of the sequence (meaning the first k terms, for some k), A occurs in at least half of the terms. We will also refer to such a sequence as A-*dominated* in order to specify that the dominating element is A.

B25   Classify each of the 16 bit strings of length 4 as either 0-dominated, 1-dominated, or not consistently dominated.

B26   List all of the A-dominated sequences of length 6 that include three As and three Bs.

Problem B26 shows that of the $\binom{6}{3} = 20$ rearrangements of the string $AAABBB$, five are A-dominated. We want to count these in a systematic way. The key to doing so is to focus on the rearrangements of $AAABBB$ that are *not A-dominated*. We will see that these can be put into a one-to-one correspondence with *all* of the sequences of length 6 that include four As and two Bs.

Algorithm   For each rearrangement of $AAAABB$, find the smallest initial segment in which the As outnumber the Bs, and change all elements in that segment (As to Bs, Bs to As). Leave the rest of the sequence unchanged.

So, for example, $AAAABB$ changes to $BAAABB$; $BAAAAB$ changes to $ABBAAB$.

B27*   Fill in the following blanks.

(a) *BBAAAA* changes to _____.

(b) *ABAABA* changes to _____.

(c) _____ changes to *ABBBAA*.

(d) _____ changes to *ABABBA*.

The preceding algorithm establishes a one-to-one correspondence between all $\binom{6}{2}$ = 15 rearrangements of *AAAABB* and the rearrangements of *AAABBB* that are not *A*-dominated. Knowing this allows us to determine that the number of *A*-dominated rearrangements of *AAABBB* is 20 − 15 = 5 without having to list these sequences individually as we did in problem B26.

Standard Problem #7

Find the number of 0-dominated bit strings that contain a given number of 0s and 1s.

Let *m* be the number of 0s, let *n* be the number of 1s, and assume that $m \geq n$. Then the answer is

$$\binom{m + n}{m} - \binom{m + n}{m + 1}$$

or, equivalently,

$$\binom{m + n}{n} - \binom{m + n}{n - 1}$$

The explanation is provided by the process illustrated in the algorithm for the case $m = n = 3$. The bit strings that are not 0-dominated are in a one-to-one correspondence with all of the bit strings of length $m + n$ that include $m + 1$ 0s and $n - 1$ 1s. The number of these strings is then subtracted from the total number of bit strings that consist of *m* 0s and *n* 1s.

B28 Find the number of 0-dominated rearrangements of the string 0011100.

B29 In this problem, we want to count the paths that go from point A to point B in the following diagram, moving along the lines and in each step, moving either one unit down or one unit to the right.

(a) The number of paths is $\binom{x}{y}$, where $x = $ _____ and $y = $ _____ (fill in the blanks).

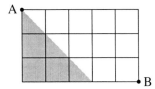

(b) How many paths are there that include no two consecutive downward moves?

(c) How many paths are there that do not enter the shaded region?

Section Summary

A *combination* is a list in which the order of the elements is unimportant. In general, a combination can contain repeated elements. The number of combinations consisting of k distinct elements from an n-element set is given by the *combination number*

$$\binom{n}{k} = \frac{n(n-1)(n-2)\cdots(n-k+1)}{k!} = \frac{n!}{k!(n-k)!}$$

The combination numbers appear on Pascal's triangle, and satisfy the relation

$$\binom{n}{k} + \binom{n}{k+1} = \binom{n+1}{k+1}.$$

The combination number $\binom{n}{k}$ also counts the rearrangements of k As and $n-k$ Bs, and consequently it is the coefficient of $A^k B^{n-k}$ in the expansion of $(A+B)^n$. Similarly, $\binom{m+n}{n}$ counts the rearrangements of m 0s and n 1s. In a variation on this, $\binom{m+1}{n}$ counts the rearrangements of m 0's and n 1's that contain no two consecutive 1s.

The number

$$\binom{n}{k}_R = \binom{n+k-1}{k}$$

counts the k-element combinations from an n-element set in which repeated elements are allowed, whereas $\binom{k-1}{n-1}$ counts those combinations in which repeated elements are allowed and each of the n elements occurs at least once.

In a *0-dominated sequence* of m 0s and n 1s, each initial segment contains at least as many 0s as 1s. The number of such sequences is

$$\binom{m+n}{n} - \binom{m+n}{n-1}.$$

More Problems

B30 Extend Pascal's triangle to include row 12.

B31 Find something that can be done in a thousand and one ways. (*Hint:* Look for 1001 on Pascal's triangle.)

B32 Simplify each of the following ratios.

(a) $\dfrac{\binom{n}{k}}{\binom{n-1}{k-1}}$ 　　　 (b) $\dfrac{\binom{n}{k}}{\binom{n}{k-1}}$ 　　　 (c) $\dfrac{\binom{n}{k}}{\binom{n-1}{k}}$

B33 Solve problem A40 again, this time by counting combinations.

B34 Solve problem A41 again, this time by counting combinations.

B35 (a) From a set of five pairs of shoes, three shoes are selected at random. Find the probability that they include a pair.

(b) From five pairs of shoes, four shoes are selected at random. Find the probability that they include two pairs.

B36 (a) Five coins are flipped simultaneously. Find the probability that exactly two land "heads."

(b) Seven coins are flipped. Find the probability that three land on one side, four on the other.

B37 Prove that for each positive integer n, $(2n)!$ is divisible by $(n!)^2$. (*Hint:* Show that the ratio of these two numbers must be an integer.)

B38 Find x and y such that

$$\binom{7}{5} + 10\binom{7}{4} + \binom{10}{2}\binom{7}{3} + \binom{10}{3}\binom{7}{2} + \binom{10}{4}7 + \binom{10}{5} = \binom{x}{y}$$

Explain this combinatorially. (*Hint:* Select from a set that consists of two types of elements.)

B39 Find x and y such that

$$\binom{n}{0}^2 + \binom{n}{1}^2 + \binom{n}{2}^2 + \cdots + \binom{n}{n}^2 = \binom{x}{y}$$

B40 Find the coefficient of $X^8 Y^5$ in the expansion of each of the following.

(a) $(X + Y)^{13}$

(b) $(2X - Y)^{13}$

B41 Find the coefficient of X^6 in the expansion of each of the following.

(a) $(X + 2)^{10}$

(b) $(X^3-1)^7$

B42 Find x and y such that

$$\binom{100}{0} + 2\binom{100}{1} + 4\binom{100}{2} + \cdots + 2^{100}\binom{100}{100} = x^y$$

B43 Find the number of eight-letter words that use letters from the set $\{A, B, C\}$ and contain exactly three As. (*Hint:* Begin by selecting positions for the As.)

B44 A poker hand consists of five cards selected at random from a standard 52-card deck. Find the number of poker hands of each of the following types.

(a) Flush: five cards from the same suit.

(b) Straight flush: five consecutive cards from the same suit, allowing an ace to be counted as either high (after king) or low (as a 1).

(c) Four of a kind, plus a fifth card.

(d) Full house: three of one kind and two of another.

(e) Three of a kind, but not a full house or four of a kind.

(f) Straight: five consecutive cards, not necessarily from one suit.

B45 Calculate the probability associated with each poker hand in problem B44.

B46 In a lottery, six distinct numbers are selected at random from the set $\{1, \ldots, 50\}$ and designated as winning numbers. A player selects six distinct numbers in advance, hoping to include as many winners as possible. Find the probability that the player selects exactly k winning numbers, for each k from 0 to 6.

B47 From a standard deck of 52 cards, 13 cards are selected. Find the probability that they include at least three cards from each suit.

B48 How many of the words counted in problem B43 contain no two consecutive As?

B49 Find the number of ways to select ten distinct letters from the alphabet $\{A, B, C, \ldots, Z\}$ if no two consecutive letters can be selected.

B50 Four distinct digits are selected at random from the set $\{0, 1, 2, \ldots, 9\}$. Find the probability that the selection includes two consecutive digits.

B51   Find the number of ten-letter combinations (allowing repetition) from the set $\{A, B, C\}$ that contain:

(a)   at least one of each letter;

(b)   at least two of each letter;

(c)   at least one A, two Bs, and three Cs;

(d)   any number of each letter;

(e)   at least one A and at least two Bs.

B52   Find the number of 50-digit combinations from the set $\{0, 1, 2, 3, 4, 5\}$ in which each digit i in the set occurs at least i times in the combination.

B53   (a)   Compare the values $\binom{100}{200}_R$ and $\binom{200}{100}_R$. Which is larger?

(b)   Answer the same question for $\binom{101}{200}_R$ and $\binom{201}{100}_R$.

B54   By considering rearrangements of m 0s and n 1s, find a combinatorial explanation for the relation

$$\binom{m+1}{n}_R = \binom{m+n}{n}$$

B55   (a)   Find the number of ways to arrange m dogs and n cats in a row, so that the cats are all separated from each other. (Test your result in the case $n = 1$. It should give $(m + 1)!$.)

(b)   If m dogs and n cats are arranged in a row at random, find the probability that the cats are all separated.

B56   (a)   Find the number of ways to seat m dogs and n cats around a circular table so that the cats are all separated from each other. (Test your result in the case $n = 1$. It should give $m!$.)

(b)   If m dogs and n cats are seated at random around a circular table, find the probability that the cats are all separated.

B57   (a)   If a circular arrangement of two 0s and two 1s is selected at random, what is the probability that the 1s are separated?

(b)   Compare the answer found in (a) with that of problem B56(b) in the case $m = n = 2$. Should they be the same? What is going on here?

B58   Find the number of A-dominated sequences consisting of 11 As and 5 Bs.

B59   Suppose that m 0s and n 1s are arranged in random order, with each possible arrangement being equally likely. Show that the probability that the resulting sequence is 0-dominated is $1 - \dfrac{n}{m+1}$. (*Suggestion:* The result of problem B32(b) is helpful here.)

B60 How many 0s and how many 1s should be included in a bit string of length 10 in order to maximize the number of 0-dominated rearrangements of the string?

B61 Find a formula for the number of 0-dominated sequences of 0s and 1s having a given length L. (The formula should depend on L only, not on the number of 0s and 1s in the sequence.)

B62 An election takes place between two candidates. Candidate A wins by a vote of 1032 to 971. If the votes are counted one at a time and in a random order, determine the probability that the winner was never behind at any point in the counting. Exactly what events are you assuming to be equally likely?

B63 The integers $1, 2, 3, \ldots, 10$ are separated at random into two subsets of five numbers each and placed in two rows, with the numbers in increasing order in each row, as follows.

$$a_1 < a_2 < a_3 < a_4 < a_5$$

$$b_1 < b_2 < b_3 < b_4 < b_5$$

Find the probability that every number in the second row is greater than the number directly above it. (*Hint:* Each arrangement of numbers in two rows corresponds to a sequence of As and Bs. For example,

$$1 \quad 2 \quad 4 \quad 8 \quad 9$$
$$3 \quad 5 \quad 6 \quad 7 \quad 10$$

corresponds to the sequence $AABABBBAAB$. What is the connection, and what property would we want the sequence to have?)

B64 (a) Find x and y such that

$$\binom{k}{k} + \binom{k+1}{k} + \binom{k+2}{k} + \cdots + \binom{n}{k} = \binom{x}{y}$$

by looking at examples on Pascal's triangle (x and y depend on k and n).

(b) Find a combinatorial explanation for the result observed in (a).

B65 (a) Show that $m^2 = m + 2\binom{m}{2}$ for all positive integers m. (*Note:* $\binom{n}{k} = 0$ if $n < k$, so $\binom{1}{2} = 0$.)

(b) Use (a) along with problem B64 to obtain a formula for

$$1^2 + 2^2 + 3^2 + \cdots + n^2 = \sum_{m=1}^{n} m^2$$

B66 Use the method of the previous problem to obtain a formula for

$$1^3 + 2^3 + 3^3 + \cdots + n^3 = \sum_{m=1}^{n} m^3$$

B67 Prove that if n is a prime integer and $1 \leq k < n$, then $\binom{n}{k}$ is divisible by n.

B68 Let n and k be relatively prime positive integers (which means that they have no common factor greater than 1). Show that

(a) $\binom{n-1}{k-1}$ is divisible by k. (*Hint:* What is: $\frac{n}{k}\binom{n-1}{k-1}$?)

(b) $\binom{n}{k}$ is divisible by n.

(c) $\binom{n}{k}_R$ is divisible by n.

B69 Show that $\binom{2n}{n}$ is divisible by $n + 1$.

B70 (a) Prove algebraically that for any non-negative integers $n \geq m \geq k$,

$$\binom{n}{m}\binom{m}{k} = \binom{n}{k}\binom{n-k}{m-k}$$

(b) Find a combinatorial explanation for the result in (a).

B71 Use problem B70 to evaluate each of the following sums.

(a) $\displaystyle\sum_{i=0}^{20} \binom{50}{i}\binom{50-i}{20-i}$

(b) $\displaystyle\sum_{i=0}^{20} (-1)^i \binom{50}{i}\binom{50-i}{20-i}$

(c) $\displaystyle\sum_{i=20}^{50} \binom{50}{i}\binom{i}{20}$

(d) $\displaystyle\sum_{i=20}^{50} (-1)^i \binom{50}{i}\binom{i}{20}$

B72 Establish the formula

$$\binom{100}{1} + 2\binom{100}{2} + 3\binom{100}{3} + \cdots + 100\binom{100}{100} = 100(2^{99})$$

in three different ways.

(a) By using the symmetry property $\binom{100}{k} = \binom{100}{100-k}$.

(b) By using the fact that $\binom{100}{k} = \frac{100}{k}\binom{99}{k-1}$ for $k \geq 1$.

(c) By differentiating the expansion of $(1 + x)^{100}$.

B73 Evaluate the sum

$$\binom{100}{0} + \frac{1}{2}\binom{100}{1} + \frac{1}{3}\binom{100}{2} + \cdots + \frac{1}{101}\binom{100}{100}.$$

C

Distributions

Suppose that we have a set of objects that are to be distributed to a number of different locations. Each object goes to one location. We can think of this as putting balls into boxes. The resulting assignment of objects to locations, or balls to boxes, is called a *distribution*.

Example   Five balls, numbered 1 through 5, are distributed into three boxes (A, B, C). One distribution is shown in the following figure.

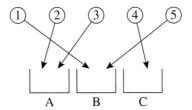

To determine the total number of distributions of five balls into three boxes, we consider placing the balls one at a time. Ball 1 can go into any one of the three boxes, and for each placement of ball 1 there are three ways to place ball 2. This gives nine ways to distribute balls 1 and 2 into the boxes. Continuing in this way, we count a total of 3^5 ways to distribute all five balls.

C1* This result is the same as the number of five-letter words that use letters from a three-letter set. What is the connection? Specifically, what five-letter word corresponds to the distribution shown in the preceding example?

C2 State a distribution problem whose answer is 5^3. What would the equivalent word-counting problem be?

C3 Find the number of ways to put ten people into five rooms.

C4 Find the number of ways to distribute 52 cards to four people. (It is not assumed that each person receives an equal number of cards.)

C5 Find the number of functions from a seven-element set to a ten-element set. (*Hint:* Think of one of the sets as a set of balls and the other as a set of boxes. Which is which?)

Standard Problem #8
Find the number of distributions of a given set of distinct balls into a given set of distinct boxes. Equivalently, find the number of functions from one set to another.

The answer is n^m, where m is the number of balls and n is the number of boxes. This also gives the number of functions from an m-element set to an n-element set.

C6* Find the number of ways to distribute five balls into eight boxes if at most one ball can go into each box.

C7 Find the number of one-to-one functions from an m-element set into an n-element set. (Assume $m \leq n$.)

C8* Find the number of distributions of five balls into three boxes if no box is allowed to be empty. (*Hint:* First count the number of forbidden distributions.)

C9 Explain the connection between problem C8 and problem A11.

C10 Find the number of functions from a five-element set *onto* a three-element set.

Distributions of Identical Objects

Suppose we erase the numbers from the balls that are to be distributed. In other words, we are now distributing identical balls into distinct boxes. Previously, we considered all balls and all boxes to be distinct.

Example   Five identical balls are distributed into three distinct boxes. All that matters is how many balls go into each box; it doesn't matter which balls they are.

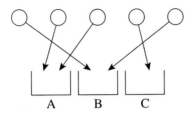

The distribution in this example can be described by the word *AABBC*, or by any rearrangement of this word. So it is actually the *combination* of letters that describes the distribution. The distributions of five identical balls into three distinct boxes correspond to the five-letter combinations, allowing repetition, from the set $\{A, B, C\}$. We know from Standard Problem #6 that the number of these combinations is

$$\binom{3}{5}_R = \binom{7}{5} = \binom{7}{2} = 21$$

Standard Problem #9
Find the number of distributions of a given set of identical balls into a given set of distinct boxes.

The answer to Standard Problem #9 is

$$\binom{n}{m}_R = \binom{n + m - 1}{m}$$

where m is the number of balls and n is the number of boxes.

As the preceding example suggests, the explanation of this result is that each distribution can be described by a combination of letters, where each letter represents a different box. The number of times each letter occurs in the combination is determined by the number of balls in the corresponding box.

C11   How many ways can 52 identical cards be distributed to four distinct people?

C12*   Find the number of ways to distribute five red balls and five blue balls into three distinct boxes. (*Hint:* Start with one color.)

C13   Find the number of ways to distribute three red, four blue, and five green balls into four distinct boxes.

C14   Find the number of ways to distribute five identical balls into three distinct boxes if each box must contain at least one ball.

Standard Problem #10
Find the number of distributions of a given set of identical balls into a given set of distinct boxes, if no boxes are allowed to be empty.

The answer to Standard Problem #10 is $\binom{m-1}{n-1}$, where m is the number of balls and n is the number of boxes. These distributions correspond to combinations with no missing elements, so the result follows from Standard Problem #5.

C15*   Find the number of distributions of 52 identical cards to four distinct people if each person must receive at least five cards.

C16   Find the number of distributions of 25 identical balls into boxes A, B, and C if box A must contain at least three balls, box B must contain at least five balls, and box C must contain *exactly* six balls.

Distribution Numbers

We return to the problem of counting distributions of distinct objects. Now, however, we specify how many objects go to each location.

C17   Find the number of ways to distribute 10 distinct balls into boxes A, B, and C if exactly five balls go to box A, exactly three balls go to box B, and exactly two balls go to box C. (*Hint:* First decide which five balls go to box A. Then decide which three go to box B.)

C18*   Find the number of rearrangements of the word *AAAAABBBCC*. (*Hint:* What does this have to do with the preceding problem?)

C19   Generalize problem C17 as follows. Distribute $a + b + c$ distinct balls into boxes A, B, and C such that a balls go to box A, b balls go to box B, and c balls go to box C, and show that the number of ways to do this is equal to

$$\frac{(a + b + c)!}{a!\,b!\,c!}$$

C20   Find the number of rearrangements of the 300-letter word *ABCABC ... ABC*.

Problem C19 generalizes in an obvious way to distributions into any number of boxes.

Standard Problem #11
Find the number of distributions of a given set of distinct balls into a given set of distinct boxes, if each box must contain a specified number of balls.

The answer to Standard Problem #11 is

$$\frac{m!}{m_1!\,m_2!\cdots m_n!}$$

where $\begin{cases} m = \text{number of balls} \\ n = \text{number of boxes} \\ m_1 = \text{number of balls in box 1} \\ m_2 = \text{number of balls in box 2} \\ \quad\vdots \\ m_n = \text{number of balls in box } n \end{cases}$ and we assume that $m = m_1 + \cdots + m_n$

We will refer to such a distribution as a *distribution of type* $(m_1, m_2, \ldots, m_n)$. The answer to Standard Problem #11 counts the distributions of type $(m_1, m_2, \ldots, m_n)$. This number is called a *distribution number* or, for reasons we will see later, a *multinomial coefficient*.

Notation

$$\binom{m}{m_1, m_2, \ldots, m_n} = \frac{m!}{m_1! m_2! \cdots m_n!}$$

We now explain why this value gives the correct result for distributions of type $(m_1, \ldots, m_n)$. The number of ways to select m_1 balls to go into box 1 is $\binom{m}{m_1}$. Of the remaining $m - m_1$ balls, m_2 of them must go into box 2. There are $\binom{m-m_1}{m_2}$ ways to select them. Continue in this way. The resulting number of distributions of all m balls into the n boxes is the product

$$\binom{m}{m_1}\binom{m - m_1}{m_2}\binom{m - m_1 - m_2}{m_3} \cdots \binom{m - m_1 - \cdots - m_{n-1}}{m_n}$$

C21 Use factorials to show that the product above is equal to the distribution number

$$\binom{m}{m_1, m_2, \ldots, m_n}$$

C22 Find the number of ways to distribute 52 cards to four distinct people with 13 cards going to each person, if

(a) the cards are distinct;
(b) the cards are identical.

As we have seen, counting distributions is equivalent to counting rearrangements. Consequently, we can solve the following standard problem.

Standard Problem #12
Find the number of rearrangements of a given word.

The answer to Standard Problem #12 is the distribution number

$$\binom{m}{m_1, m_2, \ldots, m_n} = \frac{m!}{m_1!\, m_2! \cdots m_n!}$$

where n is the number of distinct letters and $m_1, m_2, \ldots$ represent the number of times each letter occurs. m is the length of the word; necessarily, $m = m_1 + m_2 + \cdots + m_n$.

C23 Find the number of rearrangements of the word $AAABBBCCC \ldots ZZZ$.

C24* Find the number of rearrangements of the word $BANANA$ in which each of the following conditions is satisfied.

(a) The two Ns appear next to each other. (*Hint:* Glue them together.)
(b) Every N is followed by an A.
(c) The two Ns are separated.
(d) The three As are all separated.

Multinomial Expansions

The binomial expansions of section B can be generalized to expansions of the type

$$(A_1 + A_2 + A_3 + \cdots + A_n)^m$$

which are called *multinomial expansions*. For example, when $n = 3$ we have $(A + B + C)^m$, which consists of terms $A^{m_1} B^{m_2} C^{m_3}$, where $m_1 + m_2 + m_3 = m$. The coefficient of this term is

$$\binom{m}{m_1, m_2, m_3}.$$

Example The coefficient of A^3B in the expansion of $(A + B + C)^4$ is

$$\binom{4}{3, 1, 0} = \frac{4!}{3!\,1!\,0!} = 4.$$

C25 Find the coefficient of XY^5Z^2 in each of the following.

(a) $(X + Y + Z)^8$

(b) $(X - Y + 3Z)^8$

C26* Find the coefficient of X^2Y^3 in each of the following.

(a) $(X + Y + 1)^7$

(b) $(X^2 + Y - 1)^7$

In the more general situation of a multinomial expansion $(A_1 + A_2 + \cdots + A_n)^m$, the term $A_1^{m_1} \ldots A_n^{m_n}$ appears with the coefficient $\binom{m}{m_1, m_2, \ldots, m_n}$, assuming that $m_1 + m_2 + \cdots + m_n = m$. As in a binomial expansion, each term comes from combining all rearrangements of the same letters. The coefficient is the number of rearrangements, which we know is the distribution number.

C27 Find the number of rearrangements of each of the words $AABC$, $ABBC$, and $ABCC$. Use the result to count the number of four-letter words that use letters from the set $\{A, B, C\}$ in which each letter appears at least once (no missing letters).

C28 Use the method of problem C27 to find the number of five-letter words that use letters from $\{A, B, C\}$ with no missing letters. (*Suggestion:* Start by counting the rearrangements of $AAABC$ and $AABBC$.)

C29 Use the result of problem C28 to find the number of distributions of five distinct balls into three distinct boxes with no empty boxes.

Standard Problem #13
Find the number of distributions of a set of distinct balls into a set of distinct boxes, if no boxes can be empty.

Standard Problem #14
Find the number of words of a given length from a given set of letters, if each letter must occur at least once in each word.

Standard Problems #13 and #14 are equivalent. The answer to both is the sum of all of the distribution numbers in which

m = number of balls = length of each word
n = number of boxes = number of letters in the set

and the numbers $m_1, \ldots, m_n$ run through all possible sequences of n positive integers adding up to m:

$$\sum_{\substack{m_1 \ldots, m_n \geq 1 \\ m_1 + \cdots + m_n = m}} \binom{m}{m_1, \ldots, m_n}$$

Notation The sum above is denoted by $T(m, n)$.

Examples

$$T(3, 2) = \binom{3}{1, 2} + \binom{3}{2, 1} = 3 + 3 = 6$$

$$T(4, 2) = \binom{4}{1, 3} + \binom{4}{2, 2} + \binom{4}{3, 1} = 4 + 6 + 4 = 16$$

$$T(4, 3) = \binom{4}{1, 1, 2} + \binom{4}{1, 2, 1} + \binom{4}{2, 1, 1} = 12 + 12 + 12 = 36$$

$$T(5, 3) = 3\binom{5}{1, 1, 3} + 3\binom{5}{1, 2, 2} = 60 + 90 = 150$$

(Compare the last two results with the answers to problems C27–C29.)

C30 Calculate $T(6, 3)$ and $T(6, 4)$.

C31 Find $T(m, 1)$ and $T(m, m)$ for any m. Interpret the results in terms of Standard Problems #13 and #14.

C32* Find the number of five-letter words that use letters from the set $\{A, B, C, D, E\}$ and contain exactly three different letters. (*Hint:* Which three?)

C33 Find the number of distributions of seven distinct balls into three distinct boxes if at least two balls must go into each box.

The *T*-Number Triangle

The numbers $T(m, n)$ can be arranged on a triangle similar to Pascal's triangle. The row number is represented by m, starting with $m = 1$. The number of the diagonal column is represented by n, also starting with 1. For example, $T(5, 3) = 150$ appears in row 5 and diagonal column 3.

Notice that the numbers $T(m, 1) = 1$ are on the left end of each row, while on the right are $T(m, m) = m!$.

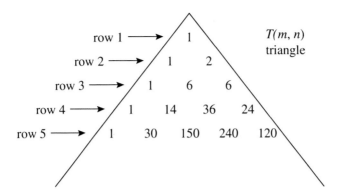

C34* Suppose we look for a pattern on the T-number triangle, such as the one that occurs on Pascal's triangle. What seems to happen? Fill in the box with the appropriate formula.

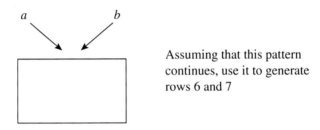

Assuming that this pattern continues, use it to generate rows 6 and 7

The observed pattern continues throughout the T-number triangle, providing a more efficient way to generate the T-numbers. Another way to express this pattern is by the relation

$$T(m, n) = n(T(m - 1, n - 1) + T(m - 1, n)) \quad \text{for } 1 < n < m$$

To see why this works, look at $T(5, 3)$ and think of it as the number of five-letter words from $\{A, B, C\}$ with no missing letters. There are three choices for the first letter. After this, the remaining four letters must be filled in, and the first letter (call it X) does not have to be used again. There are two cases.

(1) If X does *not* occur again, then the word can be completed in $T(4, 2)$ ways.

(2) If X *does* occur again, then the number of ways to complete the word is $T(4, 3)$.

We conclude that

$$T(5,3) = 3\,(T(4,2) + T(4,3))$$

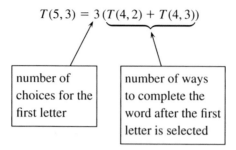

| number of choices for the first letter | number of ways to complete the word after the first letter is selected |

The same reasoning applies in general for $T(m, n)$.

Another way to calculate the T-numbers is by the following formula, which will be explained in Section E.

$$T(m,n) = \sum_{k=0}^{n}(-1)^k \binom{n}{k}(n-k)^m = n^m - n(n-1)^m + \binom{n}{2}(n-2)^m - \cdots$$

Examples

$$T(5,3) = 3^5 - 3(2^5) + 3(1^5) = 243 - 96 + 3 = 150$$

$$T(7,4) = 4^7 - 4(3^7) + 6(2^7) - 4(1^7) = 8400$$

Notice that the summation actually stops at $k = n - 1$ since the last term is always zero.

An alternative way to represent this formula is in dot product notation.

Examples

$$T(5,3) = (1,3,3,1) \cdot (3^5, -2^5, 1^5, -0^5)$$

$$T(7,4) = (1,4,6,4,1) \cdot (4^7, -3^7, 2^7, -1^7, 0^7)$$

In general,

$$T(m,n) = \left(1, n, \binom{n}{2}, \ldots, 1\right) \cdot \left(n^m, -(n-1)^m, (n-2)^m, \ldots\right)$$

| row n of | mth powers with |
| Pascal's triangle | alternating signs |

C35 Use the preceding formula to calculate $T(10,3)$ and $T(10,4)$.

Ordered Distributions: The Flagpole Problem

We consider one more variation on the problem of counting distributions. Suppose that balls are placed into narrow boxes in which they stack up vertically. Then we might want to take into account the position that each ball occupies in its box, as well as which box it goes into.

For example, the folllowing would be two different distributions of three distinct balls into two distinct boxes, taking position into account.

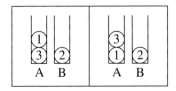

We will refer to these as *ordered distributions*.

Another interpretation of the same problem: How many ways can a set of distinct flags be placed on distinct flagpoles? Clearly the order in which the flags appear on each pole makes a difference.

C36 How many ways can three flags be placed on two poles? Two ways are shown below. Find the answer by counting directly.

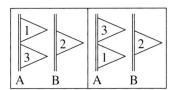

Here is a systematic way to look at problem C36. First place flag 1 on one of the poles. This can be done in two ways. Next, place flag 2, which can go on the same pole with flag 1, either above or below it, or on the other pole. So, there are three ways to place flag 2 after flag 1 has been placed.

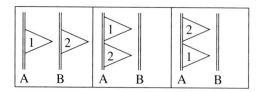

Finally, we consider where flag 3 can go. If flags 1 and 2 are on different poles, then flag 3 can be placed in four different ways, and the same is true if flags 1 and 2 are on the same pole. (Assume there is enough room above, below and between any previously placed flags.) In either case, flag 3 can be put in any of four different places. The total number of ways to place all three flags is $2 \cdot 3 \cdot 4 = 24$.

C37* Generalize this result to m flags and n poles. How many ways can the flags be placed?

C38 How many ways can m identical flags be placed on n distinct poles?

C39 Find the ratio of the answers to the two preceding problems. Can you explain this result?

Section Summary

In a *distribution*, each member of one set (think of a set of balls) is assigned to a member of a second set (think of boxes). There are n^m distributions of m distinct balls into n distinct boxes. Each such distribution is a function from the set of balls to the set of boxes, and corresponds to an m-letter word that uses letters from an n-letter set.

The number of distributions of m identical balls into n distinct boxes is $\binom{n}{m}_R = \binom{n+m-1}{m}$ if empty boxes are allowed, and $\binom{m-1}{n-1}$ if not. The *distribution number*

$$\binom{m}{m_1, m_2, \ldots, m_n} = \frac{m!}{m_1! \, m_2! \cdots m_n!}$$

counts distributions of m distinct balls into n distinct boxes with m_i balls going into the ith box, for each $i = 1, \ldots, n$. The same distribution number is the coefficient of $A_1^{m_1} \ldots A_n^{m_n}$ in the expansion of $(A_1 + \cdots + A_n)^m$, where

$m = m_1 + \cdots + m_n$, and counts the rearrangements of an m-letter word that uses letters $A_1, \ldots, A_n$, with A_i occurring exactly m_i times.

For fixed values of m and n, the sum of all distribution numbers for which $m_1 + \cdots + m_n = m$ and all $m_i \geq 1$ counts distributions with no empty boxes or, equivalently, words with no missing letters. This sum is denoted by the symbol $T(m, n)$ and satisfies the relation

$$T(m, n) = n(T(m - 1, n - 1) + T(m - 1, n))$$

for $1 < m < n$.

In an *ordered distribution*, the balls occur in a particular order within each box. Equivalently, think of placing m distinct flags on n distinct poles. The number of ways to do this can be counted by the product rule, with the result $n(n + 1) \cdots (n + m - 1)$.

More Problems

C40 Find the number of ways to distribute 18 balls, three each of six different colors, into three boxes.

C41 (a) Find the number of ways to put ten pairs of socks into four drawers if each pair is a different color and both members of a pair do not have to go into the same drawer.

(b) Change the socks to gloves. They are all the same color, but each pair contains a left and right glove.

C42 Find the number of ways that seven red balls and eight blue balls can be placed in three boxes, if

(a) each box contains at least one of each color;

(b) each box contains at least two of each color.

C43 (a) How many ways can 12 people be put into four rooms?

(b) What if each room must contain three people?

C44 Find the number of rearrangements of *AABBCCDDEE* such that each of the following conditions is satisfied.

(a) The two *A*s appear next to each other.

(b) The two *A*s are separated.

(c) The four vowels (A, A, E, E) are all separated.

C45 (a) Find the number of ways to rearrange the word $AARDVARK$.

(b) What if each R must be preceded by an A?

(c) What if no more than two As can appear together?

C46 Find the coefficient of $X^2 Y^4 Z$ in the expansion of each of the following.

(a) $(2X + Y - Z)^7$

(b) $(X + Y^2 + Z)^5$

(c) $(Y + Z - X^2 + 2)^9$

C47 The 52 cards from a standard deck are distributed to four distinct people, 13 to each. Find the probability that each person gets at least three cards from each suit.

C48 Explain, in terms of Standard Problem #13 or #14, why

$$T(m, 2) = 2^m - 2 \quad \text{and} \quad T(m, m - 1) = \binom{m}{2}(m - 1)!$$

C49 Find the number of six-letter words that use letters from a three-letter set in which each letter appears

(a) at least once;

(b) twice.

C50 (a) Find the number of one-to-one functions from a four-element set into a seven-element set.

(b) Find the number of functions from a seven-element set *onto* a four-element set.

C51 Express $T(7, 3)$ in terms of distribution numbers.

C52 Find the following.

(a) $\displaystyle\sum_{\substack{a+b+c=8 \\ a,b,c \geq 1}} \binom{8}{a,\ b,\ c}$

(b) $\displaystyle\sum_{\substack{a+b+c=8 \\ a,b,c \geq 0}} \binom{8}{a,\ b,\ c}$

C53 Repeat problem C33 with eight balls.

C54 Find the number of ways to distribute seven distinct balls into three distinct boxes if each box must contain a different number of balls, allowing an empty box.

C55 Find the number of seven-letter words that use letters from a four-letter set and in which each of the following conditions holds.

(a) All four letters appear.
(b) Exactly three different letters appear.
(c) Exactly two different letters appear.

C56 Find the probability that a five-letter word that uses letters from a six-letter set contains exactly three different letters.

C57 Find a problem whose answer is $\binom{n}{k} T(m,k)$.

C58 Find x and y such that

$$T(100,5) + 5T(100,4) + 10T(100,3) + 10T(100,2) + 5 = x^y$$

and explain this combinatorially.

C59 A card is selected at random from a standard 52-card deck. The suit (H, S, D, or C) is recorded and the card is replaced in the deck. This is done a total of seven times. Find the probability that all four suits occur among the cards selected. (*Hint:* Count sequences of suits.)

C60 In the summation formula for $T(m,n)$ that immediately precedes problem C35, suppose that $(n-k)^m$ is replaced by m^{n-k}. What is the value of the sum?

C61 Find

(a) $\displaystyle\sum_{k=0}^{100} (-1)^k \binom{100}{k} (100-k)^{100}$

(b) $\displaystyle\sum_{k=0}^{100} (-1)^k \binom{100}{k} (100-k)^{99}$

C62 Find a combinatorial explanation for the identity

$$\sum_{k=0}^{m} \binom{m}{k} T(k,n_1) T(m-k,n_2) = T(m,n_1+n_2)$$

C63 Find the number of ways to place five flags on three distinct poles if

(a) the flags are identical;
(b) the flags are distinct.

C64 Find the number of ways to place m flags on n distinct poles with at least one flag on each pole if:

(a) the flags are identical;

(b) the flags are distinct. (*Hint:* See problem C39.)

C65 Consider another approach to problem C64, in which you place all flags on one long pole and then cut it into n poles. Show that you obtain the same results.

D

Partitions

How many ways can the set $\{A, B, C, D, E, F\}$ be separated into two parts with three elements in each? For example, $ABF \mid CDE$ represents one way to do this, and it is considered the same as $CDE \mid ABF$, but $ACD \mid BEF$ is different. These are called *partitions* of the set $\{A, B, C, D, E, F\}$. What matters in a partition is the elements that are grouped together. The order of the parts and the order of the elements in each part is unimportant.

We will refer to the partitions above as *partitions of type* $(3, 3)$. One way to count them is to consider which two elements appear in the same part with A. There are $\binom{5}{2}$ ways of selecting them, and once this is done, the partition is completely determined. So the answer is 10.

D1 Find all of the partitions of $\{A, B, C, D, E, F\}$ into three two-element parts (type $(2, 2, 2)$). You should find 15 partitions.

Suppose we compare the partitions in problem D1 with distributions of the set $\{A, B, C, D, E, F\}$ into three boxes with two letters going into each box. In other words, we are comparing partitions of type $(2, 2, 2)$ with distributions of type $(2, 2, 2)$. The number of these distributions is

$$\binom{6}{2,\ 2,\ 2} = \frac{6!}{8} = 90$$

D2* Explain why the number of distributions counted in the preceding calculation is six times the number of partitions found in problem D1.

Suppose $\{A, B, C, D, E, F\}$ is separated into three unequal parts with sizes 1, 2, and 3 (a partition of type (1, 2, 3)). For example, $C \mid BE \mid ADF$ is one such partition. How many ways can this be done? First, there are six ways to decide which element is alone in one part. Then there are $\binom{5}{2}$ ways to select the elements for the two-element part. Finally, this determines the three-element part (only three elements are left). So the answer is $6\binom{5}{2} = 60$. This time, the distribution number $\binom{6}{1, 2, 3}$ gives the right answer. (Try it.)

What is going on? The explanation has to do with the sizes of the parts. In problem D2, each partition corresponded to six different distributions, since the three two-letter parts could be rearranged in $3! = 6$ ways. A partition in which all parts have different sizes will correspond to only one of the distributions counted by the appropriate distribution number.

We can state the question in a general form, as follows: If $m = m_1 + m_2 + \cdots + m_n$, how does the number of partitions of an m-element set of type $(m_1, m_2, \ldots, m_n)$ relate to the distribution number $\binom{m}{m_1, m_2, \ldots, m_n}$? The answer depends on how many of the distributions of type $(m_1, m_2, \ldots, m_n)$ correspond to the same partition.

Example Look at partitions of $\{A, B, C, D, E, F\}$ of type (1, 1, 2, 2). One of these, for example, is $A \mid F \mid BD \mid CE$. Among the distributions of the same type (1, 1, 2, 2), there are four that correspond to this same partition. For example, $A \mid F \mid CE \mid BD$ would represent a different distribution but the same partition. Therefore, the number of partitions of type (1, 1, 2, 2) is

$$\frac{1}{4}\binom{6}{1, 1, 2, 2} = \frac{6!}{16} = 45$$

D3 Find the number of partitions of a ten-element set of type (1, 1, 1, 2, 2, 3).

Standard Problem #15

For a given set, find the number of partitions of a given type.

For partitions of type $(m_1, m_2, \ldots, m_n)$, the answer is

$$\frac{\binom{m}{m_1, m_2, \ldots, m_n}}{r_1! \, r_2! \, r_3! \cdots}$$

where $m = m_1 + m_2 + \cdots + m_n$, r_1 is the number of one-element parts, r_2 is the number of two-element parts, and in general r_k is the number of

k-element parts, for each k. This is true because each partition corresponds to exactly $r_1! \, r_2! \, r_3! \cdots$ different distributions of type $(m_1, m_2, \ldots, m_n)$. These distributions are counted by the distribution number in the numerator.

D4 Find the number of partitions of the alphabet $\{A, B, \ldots, Z\}$ of type $(2, 2, 2, 3, 3, 3, 3, 4, 4)$.

D5 How many ways can a class of 25 students be divided into groups of five?

D6 Show that for any positive integer n, the number $(n^2)!$ is divisible by $(n!)^{n+1}$.

D7 Show that there are 25 partitions of the set $\{1, 2, 3, 4, 5\}$ having exactly three nonempty parts. Do this by considering the possible types separately.

D8 By comparing partitions with distributions, explain why the answer to problem D7 is $T(5, 3)/6$.

Standard Problem #16
For a given set, find the number of partitions that have a specified number of nonempty parts.

The answer to Standard Problem #16 is

$$\frac{T(m, n)}{n!}$$

where m is number of elements in the set and n is the number of nonempty parts in each partition. The explanation here is that each partition with n nonempty parts corresponds to exactly $n!$ different distributions of the given set into n boxes, with no empty boxes. These boxes are considered to be distinct when we count the distributions. However we can think of a partition as a distribution of distinct objects into identical boxes.

Notation

$$S(m, n) = \frac{T(m, n)}{n!}$$

These numbers are called *Stirling numbers of the second kind*. Thus, $S(m, n)$ counts the partitions of an m-element set with n nonempty parts.

D9 Find $S(m, 1)$ and $S(m, m)$.

The Stirling Number Triangle

Using the T-number triangle in section C, we can construct a triangle containing the Stirling numbers $S(m, n)$. As before, m is the row number and n is the diagonal column number; both start with 1.

D10*   Fill in rows 1 through 5.

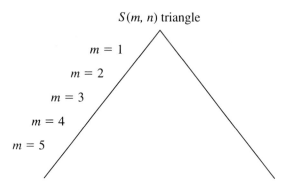

$S(m, n)$ triangle

$m = 1$

$m = 2$

$m = 3$

$m = 4$

$m = 5$

D11*   Notice that the following pattern holds for the first five rows.

Assuming that it continues, generate rows 6 through 8.

D12   Use $S(8, 5)$ to find $T(8, 5)$. Compare the result with the value of $T(8, 5)$ that would be obtained directly from the T-number triangle.

The pattern observed in problem D11 can be written as

$$S(m, n) = S(m - 1, n - 1) + nS(m - 1, n) \quad \text{for } 1 < n < m$$

D13   Establish the preceding formula by using the relation between the T-numbers

$$T(m, n) = n(T(m - 1, n - 1) + T(m - 1, n))$$

(*Suggestion:* Start with $S(m - 1, n - 1) + nS(m - 1, n)$, and convert to T-numbers.)

D14 Give a second proof based directly on what $S(m, n)$ counts. Singling out one element X of the set being partitioned, consider the following two cases:

(a) X is alone in one part.

(b) X is not alone.

D15* Find the number of partitions of a five-element set that have

(a) any number of parts;

(b) at least three nonempty parts;

(c) at most three nonempty parts.

D16 Find the number of ways to distribute five distinct balls into three identical boxes if

(a) no boxes are empty;

(b) empty boxes are allowed;

(c) at most one box is empty.

D17* Find the number of distributions of five distinct balls into two red boxes and one blue box if the two red boxes are identical and no boxes are empty. (*Hint:* First count distributions if the red boxes are distinct. How does this result compare to the one we want?)

Numerical Partitions

Next, we look at a different kind of partition. Suppose we consider all of the ways in which the number 8 can be written as a sum of three positive integers ($8 = a + b + c$), where the order of the terms a, b, c is not taken into account. We can assume that the terms appear in increasing order: $a \leq b \leq c$. Below are all of the ways to write this sum. These are called *partitions of 8 with three positive parts*.

$$
\begin{array}{l}
8 = 1 + 1 + 6 \\
8 = 1 + 2 + 5 \\
8 = 1 + 3 + 4 \\
8 = 2 + 2 + 4 \\
8 = 2 + 3 + 3
\end{array}
$$

D18   Find all of the partitions of 10 with four positive parts.

Standard Problem #17
Find the number of partitions of a given positive integer that have a specified number of positive parts.

We cannot answer this directly. Instead, we introduce a symbol to represent the answer.

Notation

$P(m, n)$ denotes the number of partitions of m with n positive parts.

So, for example, we have seen that $P(8, 3) = 5$ and $P(10, 4) = 9$. We will refer to the numbers $P(m, n)$ as *numerical partition numbers*.

D19   Find $P(m, 1)$ and $P(m, m)$.

D20   Find $P(100, 2)$ and $P(101, 2)$. In general, what is the value of $P(m, 2)$?

Like the T-numbers and Stirling numbers, the numerical partition numbers can be arranged on a triangle in which m is the row number and n is the diagonal column number, both starting at 1.

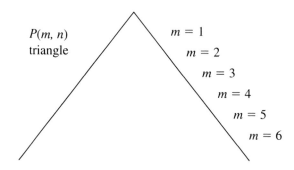

$P(m, n)$ triangle

$m = 1$
$m = 2$
$m = 3$
$m = 4$
$m = 5$
$m = 6$

D21*   Fill in rows 1 through 6. (No obvious pattern seems to hold, but we will see one later.)

Numerical partitions also can be thought of as distributions of identical objects into identical boxes. Thus $P(m, n)$ counts the number of distributions of m identical balls into n identical boxes, with no empty boxes.

D22 Find the number of ways to distribute 11 identical balls into three identical boxes, with at least two balls going into each box. (*Hint:* First put one ball in each box and then distribute the remaining balls.)

D23 Find the number of partitions of the number 11 having exactly three parts, with all parts ≥ 2.

Notation Let $P_2(m, n)$ denote the number of partitions of the number m having exactly n parts, with each part ≥ 2.

D24 Show that $P_2(m, n) = P(m - n, n)$. (*Note:* We consider $P(m, n)$ to be 0 if $m < n$.)

D25 Define $P_k(m, n)$ appropriately and find x such that $P_k(m, n) = P(x, n)$.

D26 Find the numbers $P(8, 2)$, $P_2(9, 3)$, and $P(9, 3)$. Can you explain the relationship between them?

Problem D26 suggests that, in general, $P(m, n) = P(m - 1, n - 1) + P_2(m, n)$. Combining this with the result of problem D24, we obtain the relation

$$P(m, n) = P(m - 1, n - 1) + P(m - n, n) \quad \text{for } 1 < m < n,$$

which explains how the partition numbers can be generated.

D27* Use the preceding formula to generate rows 7 through 10 of the triangle of $P(m, n)$ values.

D28 Find the number of partitions of 8 that have

(a) any number of positive parts;

(b) at least three positive parts;

(c) at most three positive parts;

(d) exactly three nonnegative parts.

D29 (a) Find $P(11, 3)$.

(b) Explain why the answers to (c) and (d) of problem D28 are equal, and why both of these are equal to $P(11, 3)$.

D30* Find x such that

$$P(m, 1) + P(m, 2) + \cdots + P(m, n) = P(x, n)$$

and explain why this is true.

Numerical Partitions with Unequal Parts

Finally, we consider partitions of a number in which the parts are unequal. For example, the partitions of 11 with three unequal positive parts are as follows.

$$
\begin{array}{|c|}
\hline
1 + 2 + 8 \\
1 + 3 + 7 \\
1 + 4 + 6 \\
2 + 3 + 6 \\
2 + 4 + 5 \\
\hline
\end{array}
$$

D31 In each of the preceding partitions, decrease the second term by 1 and the third term by 2. Use the result to explain why there are five partitions of this type.

Notation Let $P^*(m, n)$ denote the number of partitions of m having exactly n positive parts, all of which are unequal.

We saw above that $P^*(11, 3) = 5$. Also, obviously, $P^*(m, 1) = 1$.

D32 Find x such that $P^*(m, 3) = P(x, 3)$ and explain why this is true.

D33 Find x such that $P^*(m, 4) = P(x, 4)$ and explain why this is true.

D34* (a) Show that $1 + 2 + 3 + \cdots + (n - 1) = \binom{n}{2}$.

(b) Find x such that $P^*(m, n) = P(x, n)$ and explain why this is true.

D35* Find the total number of partitions of 12 having unequal positive parts.

Section Summary

In a *partition* of a set, the members of the set are separated into groups, or parts, the order of which is unimportant. A *partition of type* $(m_1, \ldots, m_n)$ contains n parts whose sizes are the m_i. The number of such partitions is given by

$$\frac{\dbinom{m}{m_1, m_2, \ldots, m_n}}{r_1!\, r_2!\, r_3! \cdots}$$

where r_k is the number of parts of size k.

The Stirling number of the second kind

$$S(m, n) = T(m, n)/n!$$

counts partitions of an m-element set with n nonempty parts and satisfies the relation

$$S(m, n) = S(m - 1, n - 1) + nS(m - 1, n)$$

for $1 < m < n$. $S(m, n)$ also counts distributions of m distinct balls into n identical boxes, with no empty boxes.

In a *numerical partition*, a positive integer is represented as a sum in which the order of the terms, or parts, is unimportant. The *partition number* $P(m, n)$ counts partitions of m with n positive parts and satisfies the relation

$$P(m, n) = P(m - 1, n - 1) + P(m - n, n)$$

for $1 < m < n$.

$$P_k(m, n) = P(m - (k - 1)n, n)$$

counts partitions of m with n parts, each of which is $\geq k$, and

$$P^*(m, n) = P(m - \tbinom{n}{2}, n)$$

counts partitions of m with n distinct positive parts.

More Problems

D36 Find the number of partitions of a 15-element set of each type below.

(a) Type $(3, 3, 3, 3, 3)$

(b) Type $(1, 2, 3, 4, 5)$

(c) Type $(2, 2, 2, 3, 3, 3)$

(d) Type $(1, 1, 1, 2, 2, 2, 3, 3)$

D37   Extend the Stirling number triangle to ten rows.

D38   Find the number of partitions of an eight-element set if

(a) all parts contain the same number of elements;

(b) each part contains an even number of elements;

(c) there must be an even number of parts. (Look for the easiest way to do this.)

D39   Find the number of partitions of the set $\{A, B, C, D, E\}$ such that

(a) A and B are in the same part;

(b) A and B are in different parts.

D40   Find the number of ways to distribute seven distinct balls into four identical boxes if

(a) no boxes are empty;

(b) at most one box is empty;

(c) no boxes are empty, and balls A and B go into different boxes.

D41   (a) Five distinct balls are distributed at random into three identical boxes. Assuming all distributions are equally likely, find the probability that no boxes are empty.

(b) Suppose the boxes are painted three different colors. Assume all distributions are equally likely, and calculate the probability that no boxes are empty.

(c) Compare results from (a) and (b). Can both of them be right?

D42   Repeat problem D17, this time allowing empty boxes. (*Hint:* Consider the different cases that correspond to which boxes are empty.)

D43   Seven distinct balls are distributed into two red boxes and three blue boxes. How many ways can this be done if

(a) no boxes are empty;

(b) empty boxes are allowed.

D44   Extend the $P(m, n)$ triangle to 12 rows.

D45 Find the following.

(a) $P_2(15, 5)$

(b) $P^*(13, 3)$

(c) $P^*(14, 4)$

D46 Find the number of distributions of ten identical balls into five identical boxes if

(a) there are no empty boxes;

(b) empty boxes are allowed.

D47 Find the number of partitions of 12 in which all parts are ≥ 2.

D48 Find the number of partitions of 15 with unequal positive parts.

D49 Find the number of partitions of 6 in which the largest part is

(a) 1   (b) 2   (c) 3   (d) 4   (e) 5   (f) 6

D50 Look at the $P(m, n)$ triangle and try to locate the sequence of answers obtained in problem D49. Guess what is true in general.

D51 Try to explain the result observed in problem D50. (*Hint:* For example, each partition of 6 that has its largest part equal to 3 can be represented by an arrangement of points, as shown in the following.)

$$3 + 3 \qquad 1 + 2 + 3 \qquad 1 + 1 + 1 + 3$$

..   ..
..   ..   .
..   .   .

D52 Find the number of partitions of 10 with positive parts if

(a) the largest part is 5;

(b) all parts are ≤ 5.

D53 Find the number of ways to place three distinct flags on two identical poles if each pole must contain at least one flag.

D54 Generalize problem D53 to m flags and n poles. Obtain a formula by using the result from problem C64(b). Check that this gives the right answer in the case $m = 3, n = 2$.

D55 Show that the same result is obtained if problem D54 is approached in the following way: Select n flags and place them at the tops of the n poles.

Then place each of the remaining flags one by one, keeping the original n flags on top.

Stirling Numbers of the First Kind

In a variation on the flagpole problem with identical poles, suppose that a set of m distinct elements is partitioned into n nonempty parts, and the members of each part are arranged around a circle. The orientation of the elements around each circle (clockwise or counterclockwise) is taken into account. For example, when $m = 3$ and $n = 1$ there are two ways to do this since all three elements are in the same part and can be arranged around a circle in two different ways.

D56 Find all of the ways to do this when $m = 4$ and $n = 2$. You should find 11 different ways.

D57 Find all of the ways to do this when $m = 4$ and $n = 3$. In this case, does it matter that the elements are arranged around a circle?

D58 When $m = 5$ and $n = 3$, the number of ways can be counted as

$$11 + 4(6) = 35$$

Explain this by singling out one element X of the set being partitioned and considering the following two cases:

(a) X is alone in one part.

(b) X is not alone.

Use the results of the two preceding problems.

In general, let $S_1(m, n)$ denote the number of ways to partition m elements into n nonempty parts and arrange the members of each part around a circle. These are the "other" Stirling numbers, the *Stirling numbers of the first kind*.

D59 Explain each of the following.

(a) $S_1(m, 1) = (m - 1)!$

(b) $S_1(m, m) = 1$

(c) $S_1(m, m - 1) = \binom{m}{2}$

D60 Use the method of problem D58 to establish the relation

$$S_1(m, n) = S_1(m - 1, n - 1) + (m - 1)S_1(m - 1, n) \quad for \ 1 < n < m.$$

D61   Use the formula in problem D60 to fill in the first six rows of a triangle containing Stirling numbers of the first kind. Start with values from problem D59.

D62   How many ways can six people be seated around three identical circular tables under each of the following conditions.

(a)   There is at least one person at each table.

(b)   Empty tables are allowed.

PART **II**

Special Counting Methods

Inclusion and Exclusion

In this section we introduce an important combinatorial method that applies in a wide variety of situations. To begin, we consider the number of elements in a union of two or more sets.

Notation For any set A, the symbol $|A|$ will represent the number of elements in A. Then, for a union of two sets,

$$|A \cup B| = |A| + |B| - |A \cap B|$$

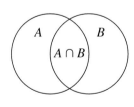 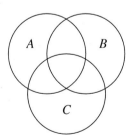

For a union of three sets,

$$|A \cup B \cup C| = |A| + |B| + |C| - \left(|A \cap B| + |A \cap C| + |B \cap C|\right) + |A \cap B \cap C|.$$

E1* Explain why the preceding formulas are correct.

The formulas can be written more easily if we simplify the notation for intersections by writing AB for $A \cap B$, ABC for $A \cap B \cap C$, etc. Then,

$$|A \cup B| = |A| + |B| - |AB|$$

$$|A \cup B \cup C| = |A| + |B| + |C| - (|AB| + |AC| + |BC|) + |ABC|$$

E2 Extend the preceding to a formula for $|A \cup B \cup C \cup D|$, the number of elements in the union of four sets.

We will use the three-set formula to calculate the number of five-letter words that use letters from the set $\{\alpha, \beta, \gamma\}$ and in which at least one of the three letters is missing. Each of the words we want to count is in at least one of the following three sets.

> A, consisting of words in which α is missing
> B, consisting of words in which β is missing
> C, consisting of words in which γ is missing.

It is important to understand that words in these sets can have more than one letter missing. For example, the word $\gamma\gamma\gamma\gamma\gamma$ is in A, since it contains no α, and it is also in B.

To count the elements in the union of the three sets, we begin by counting the elements in each set. Thinking of A as consisting of all five-letter words that use elements from the set $\{\beta, \gamma\}$, we find that $|A| = 2^5 = 32$. Similarly $|B| = |C| = 32$. The two-at-a-time intersections AB, AC and BC each contain a single element (in each case, what is it?), and the intersection ABC is the empty set (why?). Therefore,

$$|A \cup B \cup C| = 32 + 32 + 32 - (1 + 1 + 1) + 0 = 93.$$

E3 Use the preceding result to find the number of five-letter words that use letters from the three-letter set with no missing letters. (The answer should be $T(5,3) = 150$.)

E4* (a) Find the number of rearrangements of the string 1234 in which at least one of the digits is in its original position.

(b) Use the result of (a) to find the derangement number D_4.

E5* (a) Find the number of integers in the set $\{1, 2, 3, \ldots, 60\}$ that are divisible by at least one of the numbers 2, 3, and 5.

(b) Use the result of (a) to find the number of integers in $\{1, 2, 3, \ldots, 60\}$ that are relatively prime to 60. (This means they have no factors in common with 60; in other words, for each such n, the fraction $n/60$ is in lowest terms.)

The answers to problems E3, E4(b), and E5(b) can be found more directly, as follows. Let $A, B, C, \ldots$ be subsets of a larger set that includes all elements of interest in a given problem.

Notation Let $A' = U - A$, the complement of A in U.

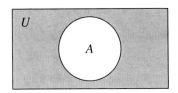

Then,

$$\left|(A \cup B)'\right| = |U| - \left(|A| + |B|\right) + |AB|$$

$$\left|(A \cup B \cup C)'\right| = |U| - \left(|A| + |B| + |C|\right) + \left(|AB| + |AC| + |BC|\right) - |ABC|$$

and a similar formula applies for any number of sets $A, B, C, \ldots$

Notice also that $(A \cup B)' = A'B'$, $(A \cup B \cup C)' = A'B'C'$, etc. Therefore, we have a formula for $|A'B'C' \ldots |$, for any number of sets.

Notation If we set

$S_0 = |U|$

$S_1 = |A| + |B| + |C| + \cdots$

$S_2 = |AB| + \cdots$ (continuing with all intersections of two sets)

$S_3 = |ABC| + \cdots$ (continuing with all intersections of three sets), etc.,

the formulas can be written

$$|A'B'| = S_0 - S_1 + S_2$$

$$|A'B'C'| = S_0 - S_1 + S_2 - S_3$$

$$|A'B'C'D'| = S_0 - S_1 + S_2 - S_3 + S_4$$

and in general, for n sets,

$$|A_1'A_2'\ldots A_n'| = S_0 - S_1 + S_2 \cdots \pm S_n = \sum_{k=0}^{n}(-1)^k S_k$$

This formula is known as *the principle of inclusion and exclusion*. As an example of how it is used, we return to problem E3. In that situation we take U to be the set of all five-letter words that use letters from the set $\{\alpha, \beta, \gamma\}$. The subsets A, B, and C of U are defined as before. In problem E3, we counted elements of U that were not in the union $A \cup B \cup C$. This is equivalent to finding the number of elements in $A'B'C'$. Using the values

$$S_0 = 3^5 = 243, \qquad S_1 = 2^5 + 2^5 + 2^5 = 96,$$

$$S_2 = 1 + 1 + 1 = 3, \qquad S_3 = 0$$

and applying the principle of inclusion and exclusion, we obtain the following result.

$$|A'B'C'| = 243 - 96 + 3 - 1 = 150$$

Once again, we have calculated $T(5, 3)$.

E6 Return to problem E4(b). What is U in this case? What are $S_0, S_1, \ldots$? Repeat the problem using the inclusion/exclusion formula.

E7 Return to problem E5(b). What is U in this case? What are $S_0, S_1, \ldots$? Repeat the problem using the inclusion/exclusion formula.

E8* Use inclusion/exclusion to find $T(m, 3)$ for any m.

E9 Use inclusion/exclusion to find the number of positive integers ≤ 100 that are not divisible by any of 2, 3, and 5.

E10* Find the number of distributions of five red balls and five blue balls into three distinct boxes, with no empty boxes.

E11 Use inclusion/exclusion to find D_5.

E12* Find the number of rearrangements of the string 11223344 that contain no two consecutive equal digits.

E13 From a standard deck of cards, the kings and queens are removed and arranged in a random order. Find the probability that there is no king and queen of the same suit next to each other. (*Hint:* Find a connection with the preceding problem.)

E14* Find the number of rearrangements of the string 12345 in which none of the sequences 12, 23, 34, and 45 occur.

E15* Find the number of five-digit strings, using digits from $\{0, 1, \ldots, 9\}$, in which there are no three consecutive equal digits. (*Hint:* Let A be the set of strings in which the first three digits are all equal.)

Combinations with Limited Repetition

Inclusion/exclusion can be used to count combinations of elements in which repetition is allowed, but with restrictions.

E16* Find the number of five-digit combinations from the set $\{1, 2, 3, 4, 5\}$, in which each of the following conditions holds.

(a) Some digit occurs at least three times.

(b) No digit occurs more than twice.

E17* Find the number of six-digit combinations from the set $\{1, 2, 3, 4, 5, 6\}$, in which no digit occurs more than twice.

E18 Suppose that you have a set of 12 colored balls, two each of six different colors C_1 through C_6. Find the number of six-ball combinations if balls of the same color are considered identical. (*Hint:* Find a connection with the preceding problem.)

E19 Find the number of nine-digit combinations from the set $\{1, 2, 3, 4\}$, in which no digit occurs more than three times.

E20 From a set of 12 colored balls, including three red, three blue, three green, and three white, nine balls are selected. How many ways can this be done? Do this by the following two methods.

(a) Find a connection with problem E19.

(b) Do it an easier way by considering the balls that are *not* selected.

Why Inclusion/Exclusion Works

To see why the principle of inclusion and exclusion produces the desired results, we return to the case of three sets.

$$|A'B'C'| = S_0 - S_1 + S_2 - S_3.$$

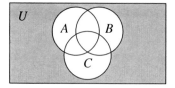

The alternating sum counts each element of U a certain number of times; an element in $A'B'C'$ (the shaded region) is counted only in $S_0(= |U|)$ and not at all in any of the other terms, so each such element contributes exactly one to the alternating sum, which is what we want it to do.

Now, look at an element in exactly one of the sets A, B, C. For example, suppose the element is in $AB'C'$. It gets counted once in S_0 and then subtracted once in S_1, and is not counted again. The net contribution of this element to the alternating sum is zero.

E21 How many times does an element in exactly two of the sets (for example, an element in ABC') get counted in each of the terms S_0, S_1, S_2, and S_3? Therefore, what is its net contribution to the alternating sum?

E22 What is the net contribution of an element in ABC to the alternating sum?

The foregoing analysis shows that each element of $A'B'C'$ contributes exactly 1 to the alternating sum $S_0 - S_1 + S_2 - S_3$, whereas all other elements contribute 0. It follows that the alternating sum is equal to the number of elements in $A'B'C'$, which establishes the inclusion/exclusion formula for three sets. In general, the same reasoning applies for any number of sets. Elements in $A_1'A_2' \ldots A_n'$ are counted once in the alternating sum $S_0 - S_1 + \cdots \pm S_n$, whereas all other elements make no contribution to this sum. This last claim depends on the fact that $1 - 2 + 1 = 0$, $1 - 3 + 3 - 1 = 0$, $1 - 4 + 6 - 4 + 1 = 0$ and, in general,

$$\sum_{k=0}^{m}(-1)^k \binom{m}{k} = 0 \quad \text{for } m \geq 1$$

(Recall why this is true; if necessary, look back at problem B14.)

E23 Let $n = 5$, and show that an element in $A_1A_2A_3A_4A_5'$ contributes 0 to the alternating sum $S_0 - S_1 + S_2 - S_3 + S_4 - S_5$.

Elements in a Given Number of Sets

The quantities $S_0, S_1, S_2, \ldots$ involved in the inclusion/exclusion formula also can be used to provide formulas for the number of elements in any given number of sets.

Example   When $n = 2$, the number of elements that are in exactly one of two sets A and B is $N_1 = |A| + |B| - 2|AB|$. This can be written as $N_1 = S_1 - 2S_2$.

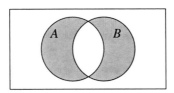

E24   Show that when $n = 3$, the number of elements in exactly one of three sets A, B, and C is given by $N_1 = S_1 - 2S_2 + 3S_3$.

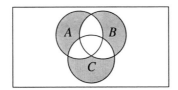

E25*   Guess the formula for the number of elements in exactly one of n sets.

E26*   (a) Use the result of problem E25 to find the number of five-letter words that use letters from a four-letter set in which exactly one letter is missing.

(b) Compare the result found in (a) with the value obtained using T-numbers.

E27   (a) Use the result of problem E25 to count rearrangements of the string 12345 having exactly one element in its original position.

(b) Explain why the answer to (a) is equal to $5D_4$.

E28   The formula for the number of elements in exactly two of n sets is

$$N_2 = S_2 - 3S_3 + 6S_4 - 10S_5 + \cdots$$

(a) Verify that this is correct for $n = 3$.

(b) Where do these coefficients appear to come from?

E29* Guess the formula for N_3.

E30 Guess the formula for N_m, the number of elements in exactly m of n sets.

Section Summary

The number of elements in a union of sets $A_1, \ldots, A_n$ is given by the alternating sum $S_1 - \cdots \pm S_n$, where the term S_k is obtained by adding the number of elements in all of the k-at-a-time intersections of the A_i. If the A_i are subsets of a set U, then the number of elements of U that are in none of the A_i is

$$|A_1' A_2' \cdots A_n'| = S_0 - S_1 + S_2 - \cdots \pm S_n = \sum_{k=0}^{n} (-1)^k S_k$$

where $S_0 = |U|$. This formula, known as *the principle of inclusion and exclusion*, has a wide variety of applications, including the calculation of $T(m, n)$ and derangement numbers, counting integers with particular divisibility properties, and counting combinations with limited repetition.

A generalization of the inclusion/exclusion formula counts elements that are in a specified number m of the given sets. The formula in the case $m = 1$ is

$$S_1 - 2S_2 + 3S_3 - \cdots \pm nS_n.$$

More Problems

E31 Use inclusion/exclusion to show that $T(7, 4) = 8400$.

E32 Use inclusion/exclusion to explain the summation formula for $T(m, n)$ given in section C:

$$T(m, n) = \sum_{k=0}^{n} (-1)^k \binom{n}{k} (n - k)^m = n^m - n(n - 1)^m + \binom{n}{2}(n - 2)^m - \cdots$$

E33 Find the number of ways to distribute three red balls, four blue balls, and five green balls into four distinct boxes with no empty boxes.

E34 Find the number of seven-letter words that use letters from the set $\{\alpha, \beta, \gamma, \delta, \epsilon\}$ and contain at least one each of α, β, and γ.

E35 Find the number of seven-letter words that use letters from the set $\{\alpha, \beta, \gamma\}$ and contain at least one α and at least two βs.

E36 Find the number of seven-letter words that use letters from the set $\{\alpha, \beta, \gamma\}$ and contain at least two of each letter. (Compare with problem C33.)

E37 Use inclusion/exclusion to explain the formula

$$D_n = \binom{n}{2}(n-2)! - \binom{n}{3}(n-3)! + \cdots = \sum_{k=2}^{n}(-1)^k\binom{n}{k}(n-k)!$$

(Why are there no terms for $k = 0$ and $k = 1$?)

E38 (a) Find $D_n/n!$, simplified as much as possible.

(b) What does this approach as $n \to \infty$?

$$\left(Hint : e^x = \sum_{k=0}^{\infty}\frac{1}{k!}x^k\right)$$

E39 The string $123\ldots n$ is rearranged in a random order. Assume that $n \geq 2$. What value of n maximizes the probability that at least one element will remain in its original position?

E40 Find the number of rearrangements of 12345 in which 1, 2, and 3 are all out of their original positions.

E41 Use inclusion/exclusion to find the number of derangements of each of the following strings.

(a) $\alpha\alpha\beta\gamma\delta$

(b) $\alpha\alpha\beta\beta\gamma\gamma$

(Compare results with those of problem A47.)

E42 Find the number of integers in the set $\{1, 2, \ldots, 210\}$ that are relatively prime to 210. (*Hint:* These are not divisible by any of 2, 3, 5, and 7.)

E43 (a) Find the number of integers in the set $\{1, 2, \ldots, 120\}$ that are divisible by at least one of 2, 3, 5, and 7.

(b) How many of the integers counted in (a) are primes?

(c) Of the integers in $\{1, 2, \ldots, 120\}$ that were *not* counted in (a), the only one which is not a prime is 1. Explain why all of the others are primes.

(d) Use the foregoing results to determine the number of primes ≤ 120.

E44 Find the number of rearrangements of the string 12345 in which none of the sequences 12, 23, 34, 45, and 51 occur.

E45   Find the number of rearrangements of the string 123456 in which none of the sequences $123, 321, 456$, and 654 occur.

E46   Find the number of rearrangements of the string 1234567 that contain at least one of the sequences $123, 234$, and 4567.

E47   Find the number of subsets of the set $\{1, 2, 3, 4, 5\}$ that do not contain any of the sets $\{1, 2, 3\}, \{2, 3, 4\}$, and $\{3, 4, 5\}$.

E48   Find the number of rearrangements of the string 111222333 that contain no three consecutive equal digits.

E49   Suppose you have a set of nine colored balls, including three red, two blue, two green, one white, and one yellow.

(a) How many ways can you select four?

(b) How many ways can you select five?

E50   Find the number of seven-digit combinations from the set $\{1, 2, 3, 4, 5\}$ if each digit can be selected at most twice. (*Hint:* Look for the easiest way to do this.)

E51   Repeat problem E50, assuming this time that each letter can be selected at most three times.

E52   Find the number of integers in the set $\{1, 2, 3, \ldots, 210\}$ that are divisible

(a) by exactly one of $2, 3, 5$, and 7;

(b) by exactly two of $2, 3, 5$, and 7.

F

Recurrence Relations

In this section, we will work with infinite sequences $(a_1, a_2, a_3, \ldots)$, usually with integer terms a_n. Sometimes it will be more convenient to have the numbering of the terms start at 0: $(a_0, a_1, a_2, \ldots)$.

A *recurrence relation* is a formula or rule by which each term of a sequence (beyond a certain point) can be determined using one or more of the earlier terms.

Examples (a) $7, 17, 27, 37 \ldots$

(b) $1, 10, 100, 1000 \ldots$

(c) $1, 3, 6, 10, 15 \ldots$

(d) $1, 2, 6, 24, 120 \ldots$

(e) $1, 1, 2, 3, 5, 8, 13 \ldots$

(f) $1, 1, 4, 10, 28, 76 \ldots$

(g) $1, 1, 1, 3, 5, 9, 17, 31 \ldots$

(h) $0, 1, 2, 9, 44, 265 \ldots$

F1 For each sequence (a)–(h), find a recurrence relation that describes the obvious (?) pattern. At what point in the sequence does the recurrence relation start to apply? Consider the first term of the sequence to be a_1.

73

F2   For sequences (a)–(d), find a nonrecursive formula for a_n. This means that a_n is expressed entirely in terms of n by a formula that does not depend on any of the other terms of the sequence.

Sequence (e) is known as the *Fibonacci sequence*. The usual notation for this sequence is

$$F_0 = 1, F_1, = 1, F_2 = 2, F_3 = 3, F_4 = 5, \ldots$$

and the recurrence relation for the Fibonacci sequence is

$$F_n = F_{n-1} + F_{n-2} \quad \text{for all } n \geq 2.$$

Sequence (h) consists of the derangement numbers D_n that appeared in sections A and E:

$$D_1 = 0, D_2 = 1, D_3 = 2, D_4 = 9, \ldots$$

Later in this section we will see that the derangement numbers satisfy two recurrence relations, as follows.

$$D_n = (n - 1)(D_{n-1} + D_{n-2}) \quad \text{for all } n \geq 3$$
$$D_n = nD_{n-1} + (-1)^n \quad \text{for all } n \geq 2.$$

We now look at a variety of combinatorial problems that can be solved using recurrence relations.

The Stamp Problem

Suppose we have 1¢, 2¢, and 5¢ stamps. The problem is to find the number of ways these can be arranged in a row so that they add up to a given value, n¢. The order of the stamps is taken into account. (Otherwise we might have used coins. That problem will be addressed in section G.) So, for example, $1 + 1 + 2$ is different from $1 + 2 + 1$. Let a_n represent the number of ways the stamps can add up to n¢. We assume that there is an unlimited supply of each type of stamp.

F3*   Find a_1, a_2, a_3, and a_4 by counting directly.

Now, we introduce a recurrence relation for the calculation of a_n in general. We consider cases according to the value of the first stamp. If this value is 1, then the total value of the remaining stamps must be $n - 1$. Therefore the number of ways in which these remaining stamps can be selected is a_{n-1}. In other words, there are a_{n-1} arrangements of stamps that have total value n and begin with a 1¢ stamp. Similarly, the number of arrangements that have total value n and begin with a 2¢ stamp is a_{n-2}. We assume that $n \geq 3$, so that the numbers a_{n-1} and a_{n-2} have already been determined. Finally, assuming that $n \geq 6$, there are a_{n-5} arrangements of stamps that have total value n and begin with a 5¢ stamp. From this analysis we arrive at the recurrence relation

$$a_n = a_{n-1} + a_{n-2} + a_{n-5} \qquad \text{for all } n \geq 6$$

F4 (a) Explain why $a_5 = a_4 + a_3 + 1$ and use this relation to find a_5.

(b) Use the recurrence relation to generate the values of a_n up to and including a_{10}.

F5* Use a recurrence relation to find the number of ways 1¢, 2¢, and 3¢ stamps can add up to 8¢.

Words with Limits on Consecutive Repetitions

In section A, we saw how we could easily count the words of a given length that use letters from a given set, in which no two consecutive letters are the same. All that was needed for that problem was the product rule. Then, in section B, we saw how to count rearrangements of a given string in which one particular symbol is not allowed to appear twice in a row. (See Standard Problem #4, in which the restriction is on the digit 1.)

Now, we will see how more complicated restrictions on consecutive repetitions can be handled using recurrence relations. To begin, consider the following problem.

Find the number of n-letter words using letters from the set $\{A, B\}$ in which there is a limit of 1 on consecutive As.

In other words, no two consecutive As can appear in the word.

F6* How does the preceding problem differ from Standard Problem #4, aside from the obvious change from digits to letters?

Notation

$w_n = $ the number of n-letter words satisfying the given condition

$a_n = $ the number of words counted by w_n that start with A

$b_n = $ the number of words counted by w_n that start with B

Obviously, $a_n + b_n = w_n$.

F7 By directly counting words, fill in the values for all $n \le 4$ in the following table.

n	1	2	3	4	5	6	7	8
a_n								
b_n								
w_n								

F8 (a) Explain why $a_n = b_{n-1}$ for all $n \ge 2$.

(b) Explain why $b_n = w_{n-1}$ for all $n \ge 2$.

(c) Use these relations to fill in the rest of the table in problem F7.

F9 Use the relations established in problem F8 to show that $w_n = w_{n-1} + w_{n-2}$ for all $n \ge 3$.

F10 What are a_n, b_n and w_n, as terms in the Fibonacci sequence?

F11* Find the number of subsets A of the set of digits $\{0, 1, 2, \ldots, 9\}$ such that A contains no two consecutive digits.

Next, we consider a variation on this problem, placing a limit of 2 on consecutive Bs while keeping the limit of 1 on consecutive As. As before, w_n counts the allowable words of length n, while a_n and b_n count the allowable words starting with A and B respectively.

F12 (a) Explain how the given conditions on consecutive equal letters are reflected in the following two tree diagrams.

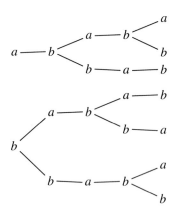

(b) Use the tree diagrams to fill in the following table.

n	1	2	3	4	5
a_n					
b_n					
w_n					

F13* (a) Explain why

$$a_n = b_{n-1} \qquad \text{for all } n \geq 2$$

and

$$b_n = a_{n-1} + a_{n-2} \quad \text{for all } n \geq 3.$$

(*Suggestion:* Think in terms of tree diagrams.)

(b) Show that

$$a_n = a_{n-2} + a_{n-3}$$
$$b_n = b_{n-2} + b_{n-3}$$

and

$$w_n = w_{n-2} + w_{n-3} \quad \text{for all } n \geq 4.$$

(c) Find w_{12}.

(*Note:* The two equations in (a) above form a *system of recurrence relations*. Although it is easy enough to derive the recurrence relations in (b), we could just as well use the system in (a) to generate further values of a_n and b_n.)

Finally, we look at one more variation on this problem. Keeping the limit of 1 on consecutive As and 2 on consecutive Bs, we include three more letters C, D, and E with no limits on consecutive occurrences. Using the notation c_n, d_n, and e_n, notice that $c_n = d_n = e_n$ for all $n \geq 1$. Also, we introduce the notation a_n', b_n', etc., to represent the number of words of a given length that do *not* begin with a particular letter. For example,

$$a_n' = w_n - a_n \quad \text{and} \quad a_{n-1}' = w_{n-1} - a_{n-1}.$$

F14 Show that

$$a_n = a_{n-1}' \qquad \text{for all } n \geq 2$$
$$b_n = b_{n-1}' + b_{n-2}' \quad \text{for all } n \geq 3$$

and

$$c_n = w_{n-1} \qquad \text{for all } n \geq 2$$

F15* Fill in the following table, using the system of recurrence relations in problem F14. Remember the Ds and Es when finding w_n.

n	1	2	3	4	5
a_n					
b_n					
c_n					
w_n					

Solving a Recurrence Relation

In certain cases, there is a method for finding an explicit formula for the nth term a_n in a sequence satisfying a recurrence relation. This method applies when the recurrence relation has the form

$$a_n = c_1 a_{n-1} + c_2 a_{n-2}$$

where c_1 and c_2 are constants (that is, they do not depend on n). The result is a formula for a_n that does not refer to earlier terms in the sequence. So, for example, the method applies to the Fibonacci sequence (where $c_1 = c_2 = 1$), but not to the derangement numbers (since $c_1 = c_2 = n - 1$ in that case).

The method consists of two steps, as indicated in the following algorithm.

Algorithm for solving a recurrence relation

$$a_n = c_1 a_{n-1} + c_2 a_{n-2}$$

given starting values a_0 and a_1:

Step 1
Find all values of r such that the geometric sequence $(1, r, r^2, \ldots, r^n, \ldots)$ satisfies the same recurrence relation.

Step 2
If two distinct r values α and β are obtained in step 1, set

$$a_n = \alpha^n A + \beta^n B;$$

If only one value $r = \alpha$ is obtained, set

$$a_n = \alpha^n(A + nB)$$

In either case, find constants A and B by using the starting values a_0 and a_1.

To be more specific, Step 1 amounts to solving the equation $r^2 = c_1 r + c_2$ for r, while in Step 2 we solve for A and B in one of the systems of equations

$$\boxed{\begin{aligned} A + B &= a_0 \\ \alpha A + \beta B &= a_1 \end{aligned}} \quad \text{or} \quad \boxed{\begin{aligned} A &= a_0 \\ \alpha(A + B) &= a_1 \end{aligned}}$$

As an example, consider the sequence $1, 1, 3, 5, 11, \ldots$, in which $a_0 = a_1 = 1$, and $a_n = a_{n-1} + 2a_{n-2}$ for $n \geq 2$.

F16 Use the algorithm to find a general formula for a_n. Check that this gives the correct result for a_7.

F17 (a) Find α and β in the case of the Fibonacci sequence. Let α represent the larger root.

(b) Without finding A and B, explain why $\alpha^n A$ is a good approximation to F_n for large values of n.

(c) What can you say about the ratio F_{n+1}/F_n as $n \to \infty$?

F18* Find A and B such that $F_n = \alpha^n A + \beta^n B$. Use a calculator to check that this gives the correct result for F_{10}. What is F_{25}?

F19 Find a formula for a_n satisfying $a_0 = 1, a_1 = 0$, and $a_n = 4(a_{n-1} - a_{n-2})$ for $n \geq 2$.

Recurrence relations of the form $a_n = c_1 a_{n-1} + c_2 a_{n-2}$, for fixed constants c_1 and c_2, are referred to as *linear with constant coefficients*. More generally, this term applies to recurrence relations of the form $a_n = c_1 a_{n-1} + \cdots + c_m a_{n-m}$, for any fixed m and constants $c_1 \ldots c_m$. The algorithm we used for problems F16–19 can be adapted to solve any linear recurrence relation with constant coefficients. Some examples in which $m = 3$ appear in problems F40 and F41.

Derangement Numbers

Earlier in this section, we observed the following two recurrence relations satisfied by the derangement numbers.

(1) $D_n = (n - 1)(D_{n-1} + D_{n-2})$ for all $n \geq 3$.

(2) $D_n = nD_{n-1} + (-1)^n$ for all $n \geq 2$.

Now, we will see why these are true, explaining the first combinatorially and then deriving the second from the first.

Looking at the case $n = 5$ to illustrate the idea, consider a derangement of the string 12345. The first digit can be 2, 3, 4, or 5. Suppose it is 2, and let $wxyz$ represent the last four digits. There are two cases:

If $w = 1$, then xyz is a derangement of 345; therefore, xyz can be chosen in D_3 ways;

If $w \neq 1$, then $wxyz$ is a derangement of 1345; therefore, $wxyz$ can be chosen in D_4 ways.

We conclude that the total number of derangements of 12345 that begin with 2 is $D_3 + D_4$.

F20 Show that the same number of derangements result from each of the other possible choices for the first digit in the derangement of 12345.

The ultimate result is that $D_5 = 4(D_3 + D_4) = 4(2 + 9) = 44$.

F21* Show by a similar argument that, in general,

$$D_n = (n - 1)(D_{n-1} + D_{n-2}) \quad \text{for all } n \geq 3$$

This explains recurrence relation (1) for the derangement numbers. To establish relation (2), set

$$a_n = D_n - nD_{n-1} \quad \text{for all } n \geq 2$$

Then, we must show that $a_n = (-1)^n$.

F22 (a) Check that $a_2 = 1$.

(b) Use recurrence relation (1) to show that $a_n = -a_{n-1}$ for all $n \geq 3$.

It follows that $a_n = (-1)^n$, and, therefore, the derangement numbers satisfy recurrence relation (2).

F23   (a) A standard deck of 52 distinct cards is shuffled. Which of the following is more likely to happen? No card is in its original position; or exactly one card is in its original position.

(b) Answer the same question in (a) if one joker is added to the deck.

Section Summary

A *recurrence relation* is a formula or rule by which each term of a sequence can be determined using one or more of the earlier terms. Examples are $F_n = F_{n-1} + F_{n-2}$ for the Fibonacci sequence and $D_n = (n-1)(D_{n-1} + D_{n-2})$ or $D_n = nD_{n-1} + (-1)^n$ for the sequence of derangement numbers. Other applications include counting the ways stamps having specified values can add up to a given total value, and counting words with limits on consecutive repetitions of particular letters.

Recurrence relations of the form $a_n = c_1 a_{n-1} + c_2 a_{n-2}$, for fixed constants c_1 and c_2, are *linear with constant coefficients* and can be solved algebraically, resulting in an explicit formula for a_n as a function of n. More general linear recurrence relations with constant coefficients $a_n = c_1 a_{n-1} + \cdots + c_m a_{n-m}$ are solvable by a generalization of the same method.

More Problems

F24   Find the number of ways a row of stamps can be worth a total of 15 cents, using stamps worth 2, 3, 5, and 7 cents each.

F25   (a) Find the number of ways a row of stamps can be worth a total of n cents, for each $n \leq 8$, if all positive integer values are allowed for the individual stamps.

(b) From the pattern observed in (a), guess what is true for all n. Can you prove it?

F26   Repeat problem F25 with the values of the individual stamps restricted to odd numbers.

F27   Let w_n count the n-letter words that use letters from $\{A, B, C, D\}$, with a limit of 1 on consecutive As and a limit of 1 on consecutive Bs. Use an appropriate system of recurrence relations to fill in the following table, in which a_n and c_n represent the number of allowable words starting with A and C respectively.

n	1	2	3	4	5
a_n					
c_n					
w_n					

F28 For each $n \leq 6$, find the number of n-letter words that use letters from $\{A, B, C\}$, if there is a limit of 1 on consecutive As and on consecutive B's, and a limit of 2 on consecutive Cs. (*Suggestion:* If necessary, look back at problems F12 through F14.)

F29 Repeat problem F28 with a limit of 3 on consecutive As and no limit on consecutive Bs and Cs.

F30 Let w_n count the n-letter words that use letters from $\{A, B, C, D, E\}$ with a limit of 2 on consecutive equal letters, and let a_n count the allowable words starting with A.

(a) Show that a_n and w_n satisfy the recurrence relations

$$a_n = 4(a_{n-1} + a_{n-2})$$
$$w_n = 4(w_{n-1} + w_{n-2}) \quad \text{for all } n \geq 3$$

(b) Find w_7. Use a calculator.

F31 Let w_n count the n-letter words that use letters from $\{A, B\}$ with a limit of k on consecutive equal letters, for some fixed positive integer k.

(a) Find a recurrence relation satisfied by w_n.

(b) Find w_{10} if $k = 3$.

F32 Generalize the result in problem F31(a), finding a recurrence relation for the number of n-letter words that use letters from an m-letter set, with a limit of k on consecutive equal letters.

F33 (a) Show that if a sequence satisfies the recurrence relation $a_n = a_{n-1} + a_{n-2} + a_{n-3}$, then it also satisfies $a_n = 2a_{n-1} - a_{n-4}$.

(b) Find a simpler recurrence relation for w_n in problem F32.

F34 Find the number of subsets A of the set of digits $\{0, 1, \ldots, 9\}$ having the property that among every four consecutive digits, A contains at least one digit and at most three.

F35 A subset of $\{1, 2, 3, \ldots, n\}$ is selected at random. If the set includes any three consecutive numbers, you win. How large must n be so that your probability of winning is greater than $\frac{1}{2}$?

F36 Find a formula for a_n satisfying $a_0 = 1, a_1 = 2$, and $a_n = a_{n-1} + 6a_{n-2}$, for all $n \geq 2$.

F37 (a) Find a formula for a_n satisfying $a_0 = 1$, $a_1 = 3$, and $a_n = 2a_{n-1} + a_{n-2}$, for all $n \geq 2$.

(b) Show that a_n is the nearest integer to $\alpha^n A$ for all $n \geq 0$, where α is the larger r value found in (a).

F38 Find a formula for a_n satisfying $a_0 = a_1 = 1$, and $a_n = -(2a_{n-1} + a_{n-2})$, for all $n \geq 2$.

F39 Find a formula for a_n satisfying $a_0 = a_1 = 1$, and $a_n = 2(a_{n-1} + a_{n-2})$, for all $n \geq 2$.

For problems F40 and F41, guess how to extend the algorithm for solving recurrence relations to apply to a recurrence relation of the form $a_n = c_1 a_{n-1} + c_2 a_{n-2} + c_3 a_{n-3}$.

F40 Find a formula for a_n satisfying $a_0 = 4$, $a_1 = -4$, $a_2 = 0$, and $a_n = 7a_{n-2} - 6a_{n-3}$, for all $n \geq 3$.

F41 Find a formula for a_n satisfying $a_0 = 0$, $a_1 = -1$, $a_2 = 2$, and $a_n = a_{n-1} + a_{n-2} - a_{n-3}$ for all $n \geq 3$.

F42 (a) Let $a_0 = 2$, $a_1 = 1$, and $a_n = a_{n-1} + a_{n-2}$ for all $n \geq 2$. Find constants A and B such that $a_n = AF_n + BF_{n+1}$ for all $n \geq 0$, where F_n and F_{n+1} are terms of the Fibonacci sequence with the usual notation.

(b) Show that any sequence satisfying the recurrence relation $a_n = a_{n-1} + a_{n-2}$ for all $n \geq 2$ can be expressed in the form $a_n = AF_n + BF_{n+1}$.

F43 Use the result of problem F10 and Standard Problem #4 to show that

$$F_n = \binom{n}{0} + \binom{n-1}{1} + \binom{n-2}{2} + \cdots \qquad \text{for all } n \geq 0$$

Counting Regions

Suppose we have n lines in a plane in general position, which means that none are parallel to each other and that no three of these lines intersect at single point.

The problem is to find the number of regions that these lines divide the plane into. Call this number r_n. Then $r_1 = 2, r_2 = 4, r_3 = 7$.

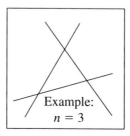

Example:
$n = 3$

F44 Find a recurrence relation for r_n. (*Hint:* How many new regions are created by the nth line?) Use it to find r_{10}.

F45 Find a nonrecursive formula for r_n. (*Hint:* see what happens when you subtract 1 from each term.)

F46 Suppose there are n circles in a plane in general position, which means that any two of the circles intersect in exactly two points, and no three circles intersect at a single point.

(a) Find a recurrence relation for r_n, the number of regions formed by these circles.

(b) Try to find a nonrecursive formula for r_n.

F47 In the algorithm for solving a recurrence relation of the form $a_n = c_1 a_{n-1} + c_2 a_{n-2}$, suppose that the equation $r^2 = c_1 r + c_2$ has two real solutions α and β or one solution α. Show that if $c_1 \neq 0$, then a_{n+1}/a_n approaches a finite limit as $n \to \infty$. What is this limit? (*Suggestion:* First show that if there are two solutions, then α and β cannot have the same absolute value, and therefore either α/β or β/α has absolute value < 1. In any case, use the result of the algorithm.)

F48 Show that the sequence that satisfies

$$a_1 = a_2 = 1 \quad \text{and} \quad a_n = (n+1)a_{n-1} - (n-1)a_{n-2} \quad \text{for all } n \geq 3$$

must also satisfy

$$a_n = na_{n-1} - 1 \quad \text{for all } n \geq 2.$$

(*Hint:* If necessary, see problem F22.)

F49 Let w_n represent the number of n-letter words that use letters from the set $\{A, B, C\}$ and contain an odd number of As. Let a_n be the number of these words that start with A.

(a) Show that

$$a_n = 3^{n-1} - w_{n-1}$$

and

$$w_n = a_n + 2w_{n-1}$$

for all $n \geq 2$.

(b) Find a recurrence relation satisfied by w_n, for all $n \geq 2$.

(c) Find a nonrecursive formula for w_n, for all $n \geq 1$.

F50 Let w_n represent the number of n-letter words that use letters from the set $\{A, B, C, D, E\}$ and contain an even number of As (possibly 0).

(a) Find a recurrence relation satisfied by w_n, for all $n \geq 2$.

(b) Calculate w_n, for all $n \leq 6$.

F51 Let S be a set of $2n$ elements and let p_n represent the number of partitions of S into n parts, with two elements in each part. Then $p_1 = 1$.

(a) Explain why $p_n = (2n - 1)p_{n-1}$, for all $n \geq 2$.

(b) Find p_5. Check your answer by using Standard Problem #15.

F52 Let S be a set of $3n$ elements and let p_n represent the number of partitions of S into n parts, with three elements in each part.

(a) Find a recurrence relation for p_n.

(b) Find p_4. Check your answer by using Standard Problem #15.

G

Generating Functions

In section E, we saw how the principle of inclusion and exclusion could be used to count combinations with limited repetition. In this section we will solve problems of this type, including more complicated variations, by a different method.

Example Find the number of ways to select a three-letter combination from the set $\{A, B, C\}$ if A can be included at most once, B at most twice, and C at most three times.

G1 Do this by counting directly.

Another approach to this problem is to form the expression

$$(1 + A)(1 + B + B^2)(1 + C + C^2 + C^3)$$

and notice that when this is multiplied out, each term represents a different combination of letters. For example, the term B^2C represents the combination *BBC*. (Notice how the term 1 in the first factor allows A to be missing from this combination.)

The terms $A^iB^jC^k$ having total degree $i + j + k = 3$ correspond to the three-letter combinations counted in problem G1. This observation suggests the following idea: If we replace A, B, and C by X everywhere they appear,

87

then the number of combinations counted in problem G1 is equal to the number of times X^3 occurs in the expansion of the product

$$(1 + X)(1 + X + X^2)(1 + X + X^2 + X^3).$$

In other words, the number of combinations is the coefficient of X^3 in this product.

Synthetic Multiplication

Multiplying polynomials can be messy and tedious, and it is easy to make mistakes. Often the same result can be accomplished more efficiently by working entirely with the coefficients, as shown in the following example.

Example Multiply $3 + X - X^2$ by $5 + 2X$, as follows.

$$
\begin{array}{rrrr}
 & 3 & 1 & -1 \\
\times & 5 & 2 & \\
\hline
15 & 5 & -5 & \\
 & 6 & 2 & -2 \\
\hline
15 & 11 & -3 & -2
\end{array}
$$

The result is $15 + 11X - 3X^2 - 2X^3$

Notice that coefficients are written beginning with the constant term and continuing with increasing powers of X. Any missing terms would be indicated by a zero. The multiplication is done from left to right with factors lined up on the left, and there is no "carrying" from one column to the next as in ordinary multiplication of integers. We will refer to this process as *synthetic multiplication*.

G2 Use synthetic multiplication to expand the product

$$(1 + X)(1 + X + X^2)(1 + X + X^2 + X^3).$$

Check that the coefficient of X^3 agrees with the number of combinations found in problem G1.

The product $(1 + X)(1 + X + X^2)(1 + X + X^2 + X^3)$ is called the *generating function* for counting combinations from a three-element set, when one element can be selected at most once, another at most twice, and the third at most three times. The coefficient of X^k in the product is equal to the number of k-element combinations that satisfy these conditions.

G3* Use the method illustrated above to construct a generating function for counting combinations without repetition from a four-element set $\{A, B, C, D\}$. Notice how this gives familiar results.

G4 Use a generating function to solve problem E16(b) again. Replace the digits with letters.

G5* From a set of nine colored balls, including three red, four blue, and two green, five balls are selected. However, it is required that at least one red ball is included and that the number of blue balls included is even. Use a generating function to find how many ways this can be done.

The Coin Problem

In section F, we counted the number of ways a sequence of stamps could add up to a given value. Now we will perform a similar procedure with coins, with the difference that the coins do not occur in any particular order. Therefore, we are counting combinations of coins adding up to a given value.

Consider the following problem. How many ways can a combination of nickels, dimes and quarters add up to 50¢? Each combination corresponds to a term of the product

$$(1 + Q + Q^2)(1 + D + D^2 + D^3 + D^4 + D^5)(1 + N + N^2 + \cdots + N^{10}).$$

G6 List all of the terms that represent 50¢. How many are there?

To solve this problem in a more systematic way, we could replace Q by X^{25}, D by X^{10}, and N by X^5 in the preceding expression and calculate the resulting product of polynomials by synthetic multiplication. The answer we want would be the coefficient of X^{50}.

G7 Why would we not want to do that?

A better idea, because all the values of these coins are multiples of 5, is to work with 5-cent units. This allows us to substitute the symbol Y in place of X^5 or, equivalently, to replace N by Y, D by Y^2, and Q by Y^5. Then our answer appears as the coefficient of Y^{10} in the product.

G8 Solve the problem in the suggested manner. Check that the answer agrees with the result found in problem G6.

G9* Suppose we have two quarters, five dimes, and ten nickels. How many combinations of these coins have a total value of 75¢? $1.00? (*Suggestion:* See whether these answers have been calculated already.)

G10 Explain why one of the answers to G9 is the same as that to G8.

Numerical Partitions Revisited

In section D, we considered the number of combinations of positive integers adding up to a given total. We took into account the number of integers in the combination, defining $P(m, n)$ as the number of ways m can be written as a sum of n positive integers, disregarding the order of the terms.

G11 Define $P(m) = P(m, 1) + P(m, 2) + \cdots + P(m, m)$. What does this represent? For example, $P(5) = 1 + 2 + 2 + 1 + 1 = 7$.

G12 Check the result of problem G11 by counting directly all of the partitions of 5.

Counting partitions of an integer is a special case of the coin problem, in which there are coins of all possible values ($1\text{¢}, 2\text{¢}, 3\text{¢}, \ldots$) and there is an unlimited supply of each type of coin. So, for example, $P(5)$ should be the coefficient of X^5 in the product

$$(1 + X + X^2 + X^3 + X^4 + X^5)(1 + X^2 + X^4)(1 + X^3)(1 + X^4)(1 + X^5)$$

G13 Check that the preceding formula gives the correct result.

G14* Define $P^*(m)$ to be the number of partitions of m having distinct positive integer terms. Construct a generating function that could be used to calculate $P^*(6)$.

G15 Find $P^*(6)$ by using the generating function found in problem G14. Check that the result is correct by directly counting partitions.

Counting Words: Exponential Generating Functions

We began this section by constructing a generating function to count combinations of letters with special conditions on the number of times each letter can appear. Now we count words rather than combinations. For example, suppose we want to count three-letter words that use at most one A, at most two Bs, and at most three Cs. Each of the three-letter combinations found in problem G1 corresponds to either 1, 3, or 6 different three-letter words.

G16 Why does each combination correspond to 1, 3, or 6 different three-letter words? Give an example of each of these situations.

G17* By considering each of the six allowable three-letter combinations found in problem G1, find the number of allowable three-letter words.

There is a way to solve such problems by working with a slightly altered generating function. To demonstrate this, we look at the following problem.

Find the number of five-letter words that use at most three As and an even number of Bs.

The usual generating function

$$(1 + A + A^2 + A^3)(1 + B^2 + B^4)$$

when expanded, contains the fifth degree terms $AB^4 + A^3B^2$. These represent two five-letter combinations. To count words, we would want these terms to appear with coefficients that indicate the number of ways each combination can be rearranged:

$$5AB^4 + 10A^3B^2$$

We could then conclude that the number of allowable five-letter words is 15. This suggests trying to change the individual factors $(1 + A + A^2 + A^3)$ and $(1 + B^2 + B^4)$ so that, when multiplied, they produce the terms $5AB^4$ and $10A^3B^2$.

Although we cannot do precisely that, what we can do will accomplish essentially the same thing and will thereby solve the problem of counting allowable words. The key to doing this is suggested by rewriting $5AB^4$ and $10A^3B^2$ as

$$\frac{5!}{1!\,4!}AB^4 + \frac{5!}{3!\,2!}A^3B^2$$

The factorials in the denominators can be built into the original factors:

$$\left(1 + \frac{A}{1!} + \frac{A^2}{2!} + \frac{A^3}{3!}\right)\left(1 + \frac{B^2}{2!} + \frac{B^4}{4!}\right)$$

When this product is expanded, the resulting fifth degree terms are

$$\left(\frac{A\,B^4}{1!\,4!} + \frac{A^3\,B^2}{3!\,2!}\right)$$

and only the factor 5! is missing. If we replace A and B with X, the product becomes

$$\left(1 + \frac{X}{1!} + \frac{X^2}{2!} + \frac{X^3}{3!}\right)\left(1 + \frac{X^2}{2!} + \frac{X^4}{4!}\right),$$

which, when expanded, contains the fifth degree term

$$\left(\frac{1}{1!\,4!} + \frac{1}{3!\,2!}\right) X^5 = \frac{1}{8}X^5.$$

From what we have seen, we expect that the number of allowable five-letter words can be obtained by multiplying this coefficient by 5! Trying this, we get 15, which is the correct result.

G18* Use the procedure illustrated above to find the number of allowable four-letter words, assuming the same restrictions on the number of As and Bs. Check your result by listing the actual words.

G19 (a) Use synthetic multiplication to expand the product

$$\left(1 + X + \frac{X^2}{2} + \frac{X^3}{6}\right)\left(1 + \frac{X^2}{2} + \frac{X^4}{24}\right).$$

(b) Write the result obtained in (a) in the form

$$1 + a_1 X + a_2\frac{X^2}{2!} + a_3\frac{X^3}{3!} + \cdots + a_7\frac{X^7}{7!}.$$

What do the numbers $a_1, a_2, \ldots, a_7$ count? For a_6 and a_7, what restriction are we assuming on the number of Bs?

The polynomial in problem G19 is called an *exponential generating function*. The name comes from the fact that the infinite series with factorial denominators

$$1 + X + \frac{X^2}{2!} + \frac{X^3}{3!} + \cdots = \sum_{k=0}^{\infty} \frac{X^k}{k!}$$

represents the exponential function e^X.

G20 Return to problem G17, and count the allowable three-letter words by constructing an appropriate exponential generating function. Start by writing

$$(1 + A)\left(1 + B + \frac{B^2}{2!}\right)\left(\qquad\qquad\right)$$

What is the third factor?

G21* For each $k \le 6$, find the number of k-letter words that use at most one A, at most two Bs, and at most three Cs.

G22* Find the number of five-letter words that use an odd number of As, at least two Bs, and at most two Cs.

G23* Return to problem G5, and assume that the order in which the balls are selected makes a difference. How many ways can five balls be selected?

Section Summary

A *generating function* is a polynomial or infinite series constructed in such a way that its coefficients provide answers to combinatorial problems. Applications include counting combinations with conditions on the number of occurrences of particular elements, and counting the ways coins having specified values can add up to a given total value. In the latter case, conditions can also be placed on the number of occurrences of each type of coin. Numerical partitions can also be counted using generating functions. In many cases calculations are simplified by the use of *synthetic multiplication*.

Exponential generating functions are used to count words with conditions on the number of occurrences of particular letters.

More Problems

G24 (a) Construct a generating function to solve problem E17. Which coefficient gives the answer?
(b) Find the answer.

G25 (a) Construct a generating function to solve problem E19. Which coefficient gives the answer? In this case, explain why either of two coefficients can be used.
(b) Find the answer.

G26 (a) Five balls, including an odd number of red balls, are selected from a set that consists of five red, three blue, two green, and one yellow ball. Construct a generating function to determine how many ways this can be done. Which coefficient gives the answer?
(b) Find the answer.

G27 (a) Construct a generating function to determine the number of seven-letter combinations from the set $\{A, B, C, D, E\}$, if A and B can be repeated any number of times but C, D, and E can be included at most once each. Which coefficient gives the answer?
(b) Find the answer.

G28 (a) Construct a generating function to determine the number of ways a combination of 1, 2, and 3-dollar bills can add up to $10. Which coefficient gives the answer?

(b) Find the answer.

G29 (a) You have an unlimited supply of three types of coins: Jefferson nickels, Buffalo nickels, and Roosevelt dimes. Construct a generating function to count the number of ways a combination of these coins can add up to 40¢. Which coefficient gives the answer?

(b) Find the answer.

G30 (a) Construct a generating function to determine the value $P(6)$. Which coefficient gives the answer?

(b) Find that coefficient and check that it gives the correct result.

G31 (a) Construct a generating function to determine the value $P^*(7)$. Which coefficient gives the answer?

(b) Find that coefficient and check that it gives the correct result.

G32 Construct a generating function to count partitions of the integer 15 in which odd terms can be included any number of times, but even terms at most once each. Which coefficient gives the answer? Don't find it.

G33 Construct a generating function to count partitions of the integer 20 in which

(a) all terms are squares;

(b) all terms are primes; (*Note:* 1 is not a prime.)

(c) the terms are distinct primes.

In each case, which coefficient gives the answer? Don't find it.

G34 (a) Construct an exponential generating function to count the five-letter words that use letters from the set $\{A, B, C, D, E\}$, in which each letter occurs at most two times. Which coefficient leads to the answer, and how?

(b) Find the answer.

G35 (a) Construct an exponential generating function to count the seven-letter words that use letters from the set $\{A, B, C, D, E\}$, in which A, B and C occur at most once each, D occurs exactly three times, and E occurs at least twice. Which coefficient leads to the answer, and how? (*Suggestion:* First think about how many Es can occur. That might simplify the procedure.)

(b) Find the answer.

G36 (a) Return to problem G26 and assume that the order of selection makes a difference. Construct the appropriate exponential generating function. Which coefficient leads to the answer, and how?

(b) Find the answer.

G37 Use an exponential generating function to find the value of $T(5, 3)$. Check that this gives the correct result.

G38 Interpret the equation

$$e^{3X} = \sum_{k=0}^{\infty} 3^k \frac{X^k}{k!}$$

in terms of counting words. (*Hint:* $e^A e^B e^C$.)

G39 Interpret the equation

$$(e^X - 1)^3 = \sum_{k=0}^{\infty} a_k \frac{X^k}{k!}$$

in terms of counting words. What is a_k?

G40 In the equation

$$(e^X - X - 1)^3 = \sum_{k=0}^{\infty} a_k \frac{X^k}{k!}$$

what does a_k count?

H

The Pólya-Redfield Method

In this section, we will develop a powerful method for solving difficult counting problems in situations that involve symmetry. We begin by showing how the method applies in a particular situation involving rotations of a string. This will lead in a natural way to a variety of other applications.

Rotations of a String

Starting with any string $X_1 \ldots X_n$, suppose we move an initial segment $X_1 \ldots X_i$ from the beginning of the string to the end. The result is another string $X_{i+1} \ldots X_n X_1 \ldots X_i$, which we will refer to as a *rotation* of the original string. The value of i can range from 0 to $n - 1$, with the rotation corresponding to $i = 0$ being the original string. For example, the rotations of the string ABC are ABC, BCA, and CAB.

Consider the 16 strings of length 4 that use letters from the set $\{A, B\}$. Some of these strings are rotations of others. For example, the four strings $AAAB$, $AABA$, $ABAA$, and $BAAA$ are all rotations of $AAAB$ (and also of each other). On the other hand, the only rotation of the string $AAAA$ is $AAAA$ itself. The 16 strings divide into subsets, called *orbits*, each of which consists of strings that are rotations of each other. One orbit is $\{AAAA\}$, consisting of a single string, while another orbit is $\{AAAB, AABA, ABAA, BAAA\}$.

The *length* of an orbit is the number of elements in the orbit. So the two orbits indicated above have lengths 1 and 4.

H1   List all of the remaining orbits in the preceding example. Including the two already listed, the orbit lengths should add up to 16.

H2*   In the set of all six-letter strings that use letters A and B, find the orbits that contain each of the strings $AAABBB$, $AABAAB$, and $ABABAB$. What do you notice about orbit lengths in relation to string length?

H3*   (a) Find the length of the orbit containing the string

$$AABCAAABCAAABCA$$

in the set of all 15-letter strings that use letters A, B, and C.

(b) Guess at a relationship that always exists between orbit length and string length in the set of all n-letter strings that use letters from a given set. How would you recognize the orbit length corresponding to a particular string?

In such situations, our main interest will be in finding the number of different orbits within a given set of strings. We will develop a procedure for doing so in which the orbit lengths play an important role.

H4   Find all of the orbits among the three-letter strings that use the letters A, B, and C. It is sufficient to list one representative from each orbit and indicate the orbit length. Check that the orbit lengths add up to the correct total.

H5   Repeat problem H4 for five-letter strings that use the letters A and B.

H6*   How many different ways can the squares in the the accompanying figure be colored, using two colors? Each square is given one color and it is not required that both colors actually occur. Consider two colorings to be the same if one coloring can be obtained from the other by rotating the figure.

H7*   How many ways can you color the edges of an equilateral triangle, using three colors? As in problem H6, consider two colorings to be the same if one can be obtained from the other by rotating the figure. Again, it is not required that all colors actually occur.

H8*   How many ways can five dogs and/or cats be seated around a circular table? (For this problem, consider two animals of the same type to be identical.)

Returning to the situation of four-letter strings that use letters A and B, we will find it useful to form the expression

$$AAAA + AAAB + AABA + ABAA + BAAA + ABAB + BABA + AABB$$

$$+ ABBA + BBAA + BAAB + ABBB + BBBA + BBAB + BABB + BBBB$$

in which each term is one of the 16 strings under consideration. Suppose that each string in this expression is replaced by the reciprocal of the length of the orbit that contains that string: $AAAA$ is replaced by 1, $AAAB$ by $\frac{1}{4}$, etc.

H9* Find the resulting sum, and explain why it is equal to the number of orbits.

Problem H9 suggests that we might be able to determine the number of orbits without actually listing all of the strings if this sum of reciprocals of orbit lengths can be generated in some other way. This is the strategy we will use. However in order to implement it, it will be necessary to introduce some additional concepts. First, we will think of rotations as transformations that can be applied to any string. Specifically, we define

$$R_i(X_1 \ldots X_n) = X_{i+1} \ldots X_n X_1 \ldots X_i$$

In other words, R_i moves the initial segment of length i to the end of the string. Then all rotations of an n-letter string are obtained by applying $R_0, R_1, \ldots, R_{n-1}$ to the string. R_0 is the identity transformation that leaves each string unchanged. The set $\{R_0, R_1, \ldots, R_{n-1}\}$ is called the *rotation group* for n-letter strings.

H10* What happens to the sum $AAAA + \cdots + BBBB$, consisting of all 16 four-letter strings that use letters A and B, when the rotation R_1 is applied to each term? Answer the same question for R_2 and R_3.

H11 Explain the following equation.

$$(R_0 + R_1 + R_2 + R_3)(AAAA + \cdots + BBBB) = 4(AAAA + \cdots + BBBB).$$

We know that if each string on the right side of the preceding equation is replaced by the reciprocal of the corresponding orbit length, the result is 4 times the number of orbits. We will use the left side of the equation to calculate the same value in a different way.

Consider what happens when $R_0 + R_1 + R_2 + R_3$ is applied to only one string. For example,

$$(R_0 + R_1 + R_2 + R_3)(AAAB) = AAAB + AABA + ABAA + BAAA$$

$$(R_0 + R_1 + R_2 + R_3)(AAAA) = 4\,(AAAA)$$

$$(R_0 + R_1 + R_2 + R_3)(ABAB) = 2\,(ABAB + BABA).$$

In each case, the result is a multiple of the orbit that contains the string. When each string in the orbit is replaced by the reciprocal of the orbit length, the result of the calculation is the coefficient (1, 4, and 2, in these examples). Therefore we want to take a closer look at these coefficients.

H12   For each of the strings $AAAB$, $AAAA$, and $ABAB$, find all members of the rotation group $\{R_0, R_1, R_2, R_3\}$ that leave the string unchanged.

For each string, the set of rotations found in problem H12 is called the *stabilizer group* of the string, and the number of rotations in this group is the *stabilizer number* of the string. Notice that the stabilizer numbers of these three strings are precisely the coefficients that appeared in the calculation preceding problem H12. Of course, this is no accident.

H13*   Try to explain the following phenomenon. In general, for a string of length n,

$$(R_0 + R_1 + \cdots + R_{n-1})(X_1 \ldots X_n) = m((X_1 \ldots X_n) + \cdots)$$

which is a multiple of the orbit containing the string, with the stabilizer number of the string as the coefficient m.

Now, we return to the problem of evaluating the expression in problem H11 when each string is replaced by the reciprocal of its orbit length. We now know that this result is equal to the sum of the stabilizer numbers of all 16 strings.

H14*   Find the sum of the stabilizer numbers of all 16 strings directly, by completing the following table.

string	AAAA	AAAB	AABA	ABAA	BAAA	ABAB	BABA	AABB
stabilizer number								

string	ABBA	BBAA	BAAB	ABBB	BBBA	BBAB	BABB	BBBB
stabilizer number								

The result obtained in problem H14 should be 4 times the number of orbits in the set of these 16 strings. Although this procedure works, it is not an efficient way to determine the number of orbits. Now, however, we take the great leap that turns this into a practical procedure: Instead of organizing the sum of stabilizer numbers by strings, we organize it by rotations. There are 16 strings, but only four rotations. For each rotation we find the number of strings that are unchanged by the rotation. For example, the identity rotation R_0 leaves all 16 strings unchanged, so the number corresponding to R_0 is 16. For R_1, R_2, and R_3, the corresponding numbers are 2, 4, and 2, respectively.

H15 Verify that the numbers 2, 4, and 2, for R_1, R_2, and R_3, are correct, by finding the strings that each one counts.

The numbers we have found are called the *invariant numbers* of the four rotations. When we add them together the result is 24, the same as the sum of the stabilizer numbers of the 16 strings.

H16* Why are the two sums equal? Explain, in terms of what they count.

We summarize what we have just observed, as follows. The sum of the invariant numbers of the rotations, $16 + 2 + 4 + 2$, is equal to the sum of the stabilizer numbers of the 16 strings, $4 + 1 + 1 + \cdots$, which in turn is equal to 4 times the sum of reciprocals of the orbit lengths of all strings. Finally, this last sum is 4 times the number of orbits. Consequently the number of orbits can be calculated as

$$\frac{16 + 2 + 4 + 2}{4}$$

In other words, the number of orbits is equal to the average of the invariant numbers of the four rotations.

H17* Use the preceding method to find the number of orbits among the five-letter strings that use the letters A and B. Compare the result with that of problem H5.

H18* Repeat problem H17 for the three-letter strings that use the letters A, B, and C. Compare the result with that of problem H4.

In general, we have described a procedure for counting orbits that are generated by rotations of strings. We will see, however, that there are other situations, not always involving strings and rotations, in which the same concepts apply and the same method counts the orbits. For the time being, we will state the procedure specifically for the situation of strings and rotations; for clarity, we will refer to these orbits as *rotational orbits*.

In the set of n-letter strings formed from a given set of letters, the number of rotational orbits is equal to

$$\frac{r_0 + r_1 + \cdots + r_{n-1}}{n},$$

where r_i is the invariant number of the rotation R_i.

H19* Use this procedure to count the rotational orbits of four-letter strings that use the letters A, B, and C.

H20 State two coloring problems that are solved by the calculation of the number of orbits in problem H19, as follows.

(a) Coloring squares in the accompanying figure

(b) Coloring the edges of a single square

In determining the invariant numbers of rotations, the following observation is often helpful.

Algorithm To find r_i when i is not a divisor of n,

Step 1 Find $d = GCD(i, n)$, the greatest common divisor of i and n;

Step 2 Find r_d.

Then $r_i = r_d$.

Example When $n = 6$ and $i = 4$, $d = GCD(4, 6) = 2$. So $r_4 = r_2$.

H21 Verify directly that $r_4 = r_2$ for six-letter strings that use the letters A and B, by finding the invariant strings for each of the corresponding rotations.

As this example suggests, the equality of r_i and r_d is based on the fact that the rotations R_i and R_d have the same set of invariant strings. A full explanation of this relationship would require concepts from number theory that are outside the scope of this book.

H22* How many ways can five pigs, ducks, and/or goats be seated around a circular table? Consider animals of the same type to be identical.

H23* Find the number of orbits among the six-letter strings that use the letters A and B.

By now, you have probably noticed that the value of r_d, when d is a divisor of n, always seems to be k^d, where k is the number of letters used. This will be true as long as all of the invariant n-letter strings that use these letters are being counted.

H24* Be sure that you understand why the preceding observation is correct.

H25* Find the invariant number of the rotation R_{10} for 15-letter strings that use the letters A, B, and C.

As a variation on the orbit-counting problem, suppose we restrict our attention to only those n-letter strings, that use a given set of k letters, in which each of the letters appears at least once. In other words, we want to count rotational orbits of strings that have no missing letters. Everything we have done still applies except for one detail: Now the calculation of the invariant numbers must take into account that all k letters occur in the string. For example, in the situation of problem H25, we would have

$$r_{10} = r_5 = T(5, 3) = 150.$$

H26* Find the number of ways to seat five pigs, ducks, and/or goats around a circular table, including at least one of each animal.

Another variation on the orbit-counting problem occurs when we allow transformations other than rotations. For example, consider again the problem of coloring four squares, allowing the figure to be flipped as well as rotated. The first two colorings indicated below are now equivalent, because the figure can be flipped around the horizontal axis running through the center; and the second coloring can be rotated into each of the next three.

A	B		C	D		A	C		B	A		D	B
C	D		A	B		B	D		D	C		C	A

Equivalently, each of the last four colorings can be obtained from the first by reflecting the figure through either a horizontal, vertical, or diagonal axis. Looking at the situation this way, we add four reflections (through four different axes: one horizontal, one vertical, and two diagonal) to the four rotations of the figure, for a total of eight transformations. We can calculate invariant numbers for the reflections and, not surprisingly, determine the number of coloring types (orbits) that use a given set of colors, by averaging all eight invariant numbers.

H27* Consider colorings of the squares in the accompanying figure that use colors A, B, and C.

(a) Find the invariant numbers of all eight transformations.

(b) Find the number of different colorings if the figure can be rotated but not flipped. (We have already done this, but do it again.)

(c) Find the number of different colorings if the figure can be flipped as well as rotated.

According to the results of problem H27(b) and (c), there must be some colorings that are equivalent when the figure is flipped but not when it is rotated.

H28 (a) Find a pair of colorings that have the property just described.

(b) Find two more pairs with the same property.

(c) Explain these results in terms of merging orbits.

H29* How many different ways can five beads be arranged on a circular loop, if each bead can be any one of three colors? Consider that the loop can be flipped as well as rotated. (*Suggestion:* First, think about applicable reflections. It might help to consider the beads to be located at the corners of a regular pentagon.)

H30* Repeat problem H29 for six beads and two colors. Compare your answer with that of problem H23, and find two colorings that explain the difference. (Again, think about applicable reflections. There are two different types.)

Now consider coloring the nine squares in the accompanying figure, using two colors, A and B. This time we allow the figure to be rotated but not flipped, so the only transformations involved are the four rotations

$$R_0 \text{ (the identity)}, R_1 \text{ (90° clockwise)}, R_2 \text{ (180° clockwise)},$$

$$\text{and } R_3 \text{ (270° clockwise)}.$$

H31* (a) The invariant numbers of these four rotations are 2^9, 2^3, 2^5 and 2^3. respectively. Explain why this is true. What is the significance of these exponents?

(b) Find the number of different colorings.

Continuing with the preceding example, suppose that we want to know how many of the colorings counted above use color A on five squares and color B on four. We can take this condition into account in determining the invariant numbers. Then r_0 becomes $\binom{9}{5}$ = 126, the number of colorings in which A occurs five times. For r_1, consider that for coloring to be unchanged by R_1, all four corner squares must be the same color, and the same holds for the four noncorner squares along the edges. It follows that the five squares colored with A must include the center square and one of these two groups of four. Therefore, $r_1 = 2$. For the same reason, $r_3 = 2$.

H32 (a) Show that $r_2 = 6$ by directly counting colorings.

(b) Find the number of colorings that use A on five squares and B on four.

The calculation of the invariant numbers can become complicated in such situations. However, there is a systematic way to approach the problem by working with a generating function.

H33 Find the coefficient of the $A^5 B^4$ term in the expansion of

$$\frac{(A + B)^9 + 2(A + B)(A^4 + B^4)^2 + (A + B)(A^2 + B^2)^4}{4}$$

The expression in problem H33 is called the *pattern inventory* for colorings of the nine-square figure using two colors. When expanded, the coefficient of each term counts the number of colorings in which each color occurs a particular number of times.

H34 Try the preceding method for colorings using one A and eight Bs. Check that the coefficient of AB^8 gives the right answer.

To understand why this works, start with the $(A + B)^9$ term. If we correspond each factor $A + B$ to one of the nine squares, then the choice of a color for a particular square corresponds to the choice of either A or B from the corresponding factor. In this way, colorings of the entire figure correspond to nine-letter strings that use A and B. The number of colorings that use A on five squares and B on four is equal to the number of strings that contain five As and four Bs and is given by the coefficient of A^5B^4 in the expansion of $(A + B)^9$.

Continuing along the same lines, consider terms in the expansion of $(A + B)(A^4 + B^4)^2$. Each of these terms comes from a choice of either A or B from the first factor, and a choice of either A^4 or B^4 from each of two more factors. These choices correspond to a choice of either A or B for the center square, a choice of either A or B for the four corner squares (all the same), and a choice of either A or B for all four remaining squares. Such a coloring is unchanged by the rotation R_1, so the coefficient of A^5B^4 in the expansion of $(A + B)(A^4 + B^4)^2$ counts colorings that use A on five squares and B on four and that are unchanged by R_1. The same applies for R_3, accounting for the coefficient 2 in the numerator of the pattern inventory.

H35* Explain why the coefficient of A^5B^4 in the expansion of $(A + B)(A^2 + B^2)^4$ counts colorings that use A on five squares, use B on four, and are unchanged by the 180° rotation R_2.

This explains why the calculation in problem H33 produces the same result as that in problem H32(b). The same reasoning applies more generally, allowing us to solve problems such as H32(b) by working with an appropriately constructed pattern inventory.

H36 (a) Decide how the pattern inventory in problem H33 should be changed to accomodate a third color C.

(b) Find the number of ways to color the nine-square figure using three colors, with each color occurring on three squares.

The construction of pattern inventories is facilitated by the introduction of polynomials that are connected with the transformations in a given problem. To illustrate this, we continue with the nine-square figure and its rotations R_0, R_1, R_2 and R_3.

Numbering the squares as shown in the accompanying figure, we consider where each square is moved by each rotation. R_1 moves square 1 to square 3, square 3 to square 9, square 9 to square 7, and square 7 to square 1. This information can be recorded more compactly by writing (1397), which we refer to as a *cycle*. At the same time, R_1 moves square 2 to square 6, 6 to 8, 8 to 4, and 4 to 2. So, a second cycle is (2684). Finally, square 5 remains unmoved, a situation denoted by (5). To summarize, the rotation R_1 has the cycle structure

(1397)(2684)(5) when applied to the nine-square figure, consisting of two cycles of length 4 and one cycle of length 1. Accordingly we assign the *cycle code* $X_4X_4X_1$, or $X_4^2X_1$, to this rotation. The factors in a cycle code can appear in any order, so the same code can be written as $X_1X_4^2$.

1	2	3
4	5	6
7	8	9

H37* Determine the cycle code for each rotation R_0, R_2, and R_3.

Next, we construct a polynomial associated with rotations of the nine-square figure by averaging the cycle codes of the four rotations.

$$\frac{X_1^9 + 2X_1X_4^2 + X_1X_2^4}{4}$$

This is known as the *cycle index polynomial* for rotations of the figure. There is also a single-variable version, obtained by setting each $X_i = X$:

$$\frac{X^9 + X^5 + 2X^3}{4}$$

To distinguish between these two polynomials, we refer to the latter simply as the *cycle polynomial*. The significance of these polynomials is demonstrated by the following problems.

H38 Set $X = 2$ in the preceding cycle polynomial. What does this count?

H39 Replace variables in the cycle index polynomial as follows.

$$X_1 = A + B, \quad X_2 = A^2 + B^2, \quad X_4 = A^4 + B^4.$$

What does this produce?

We have illustrated a procedure that can be employed more generally in situations where objects (which can be parts of a geometrical figure, beads on a loop, or positions in a string, to name just a few possibilities) are assigned "colors," which can be members of any set, and where there is a specified set of transformations that rearrange the objects and thereby determine which colorings are considered equivalent. We assume that the transformations conform to certain requirements; in particular, they must form a *group*. (See "Groups of

Transformations" on p. 114, and problems H68–70.) In that case, the method allows us to count the different colorings of the objects or, if we prefer, only those colorings that satisfy specified conditions. The details of the method are summarized below.

Cycle codes: Each transformation rearranges the objects in cycles. A cycle of length i is represented by a factor X_i in the cycle code of the transformation.

Cycle index polynomial: This is the sum of all cycle codes, one for each transformation, divided by the number of transformations. In other words, it is the average of all cycle codes.

Cycle polynomial: This is the result of setting all $X_i = X$ in the cycle index polynomial.

Pattern inventory: This is the result of setting $X_i = A_1^i + \cdots + A_k^i$ in the cycle index polynomial for each i, where the A_j are symbols representing the different colors.

Number of colorings: This is obtained by setting $X = k$ in the cycle polynomial, where k is the number of colors.

Number of colorings that use each color a specified number of times: This is given by the appropriate coefficient in the expansion of the pattern inventory.

H40* (a) Construct the cycle index polynomial for rotations of the accompanying figure, considering the four squares to be the objects being rearranged.

(b) Construct the pattern inventory for colorings of the four squares that use colors A and B and allow rotations of the figure.

(c) Predict the coefficient of each term in the expansion of the pattern inventory by directly counting colorings, and check that the results are as expected.

H41* Repeat problem H40(a), allowing reflections as well as rotations.

H42 Use results from the two preceding problems to count colorings of the four squares using three colors, both with and without reflections. Compare answers with those found in problem H27.

H43*   (a) Solve problem H30 again by constructing the appropriate polynomial.

(b) Use a pattern inventory to count the number of ways three blue beads and three white beads can be arranged on a circular loop that can be rotated or flipped. Check your answer by directly counting arrangements.

Rotations of a Cube

As a final application of the counting method presented in this section, consider the problem of coloring the six faces of a cube. Each face is given one color, and two colorings are considered the same if one can be obtained from the other by rotating the cube. Rotations, however, take place in three dimensions and are consequently more complicated than rotations in two dimensions.

Consider all of the different positions in which the cube can be placed. The front square can moved to any of six positions: front, back, top, bottom, left, or right. Then, for each of these six choices, the cube can be turned four ways. This gives 24 different positions for the cube.

Looking at it another way, think of rotating the cube around an axis that runs through the centers of the top and bottom squares. The cube can be left in its original position (the identity transformation), rotated 90° in either direction around this axis, or rotated 180°. There are similar 90° and 180° rotations around an axis running through the front and back, and around an axis running through the left and right sides. So far, then, we have ten transformations of the cube: the identity, six 90° rotations, and three 180° rotations. The cycle codes for these transformations are as follows.

$$X_1^6 \text{ for the identity transformation}$$

$$X_1^2 X_4 \text{ for each 90° rotation}$$

$$X_1^2 X_2^2 \text{ for each 180° rotation}$$

H44   Be sure you agree with these cycle codes.

Now, consider an axis that runs through two opposite corners of the cube, passing through the center. The cube can be rotated 120° in either direction around this axis. Considering all eight corners of the cube, we see that there are four such axes, resulting in eight 120° rotations.

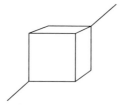

H45   Find the cycle code for each 120° rotation.

So far, we have accounted for 18 rotations of the cube. The remaining six come from considering axes that run through the midpoints of two opposite edges of the cube, such as the top left edge and the bottom right edge. Considering all pairs of opposite edges, we find six such axes, and around each one the cube can be rotated 180°.

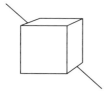

H46   Find the cycle code for each of these 180° rotations.

H47*   Construct the cycle index polynomial for the 24 rotations of the cube.

H48*   Find the number of ways to color the faces of a cube using

(a) two colors;

(b) three colors.

H49   Use pattern inventories to find the number of ways to color the faces of a cube using

(a) two colors, each on three faces;

(b) three colors, each on two faces.

Historical Note   The counting method presented in this section became widely known when it was used by George Pólya in a 1937 paper on counting chemical compounds. Later it was discovered that the same technique had appeared ten years earlier in a paper by J. H. Redfield. The basic idea of using invariant numbers to count orbits goes back even further, at least to a 1911 publication by W. Burnside. Pólya and Redfield appear to have been the first to apply the method to solving combinatorial problems.

Section Summary

A *rotation* of a string is the result of moving an initial segment from the beginning of the string to the end. An *orbit* consists of strings within a given set that are rotations of each other. The *invariant number* of a rotation (considered as a transformation that can be applied to any string) is the number of strings in the given set that are unchanged by that rotation. The orbits can be counted by averaging the invariant numbers of all rotations.

The procedure outlined above can be applied in more general circumstances whenever objects (such as parts of a geometrical figure) are assigned "colors," which can be members of any set, and where there is a specified set of transformations that rearrange the objects and thereby determine which colorings are considered equivalent. Each transformation rearranges the objects in cycles that determine the *cycle code* of the transformation. The average of all of these cycle codes is the *cycle index polynomial*, which leads to the construction of the *pattern inventory* for coloring the objects using the specified set of colors. Coefficients in the expansion of the pattern inventory indicate the number of colorings that use each color a specified number of times.

More Problems

H50 In the set of all six-letter strings that use the letters A and B, find one representative from each rotational orbit and indicate the orbit length. Check that the orbit lengths add up to the correct total.

H51 (a) Express $(R_0 + R_1 + R_2 + R_3 + R_4 + R_5)(ABAABA)$ in the form of problem H13. Verify that the coefficient in the resulting expression is equal to the stabilizer number of the string by listing the members of the stabilizer group.

(b) Repeat (a) for the string $ABABAB$.

H52 For 12-letter strings that use the letters A, B, and C, find the invariant number of the rotation R_4, taking into account

(a) all strings;

(b) only strings that contain at least one of each letter;

(c) strings that contain at least one A;

(d) strings that contain six As, three Bs, and three Cs;

(e) strings that contain four of each letter.

H53 What other rotation, if any, has the same invariant numbers as those found in problem H52?

H54 (a) Find the number of rotational orbits of seven-letter strings that use the letters A and B.

(b) Find the number of ways to arrange seven beads of two colors on a circular loop that can be rotated or flipped.

(c) Find examples that explain the difference between your answers to (a) and (b).

H55 Repeat (a) and (b) of problem H54, using eight letters or beads.

H56 Repeat problem H27(c) for colorings in which each of the three colors occurs on at least one square.

H57 (a) Find the number of rotational orbits of six-letter strings that use the letters A, B, and C.

(b) How many of the strings counted in (a) contain at least one of each letter?

H58 (a) Find the number of ways to arrange six beads of three colors on a circular loop that can be rotated or flipped.

(b) How many of the arrangements counted in (a) contain at least one bead of each color?

H59 (a) Use a pattern inventory to find the number of ways to arrange seven beads—one red, two blue and four green—on a circular loop that can be rotated or flipped.

(b) Suppose red beads cost 2¢ each, blue beads 3¢, and green beads 5¢. Figure out a way to use the pattern inventory constructed in (a) to count the arrangements of seven red, blue, and green beads having a given total cost. (*Suggestion:* If necessary, look back at the *coin problem* in section G.)

H60 (a) Modify the cycle index polynomial given after problem H37 to take into account reflections of the figure as well as rotations.

(b) Find the number of colorings of the nine squares using color A on five squares and B on four, if the figure can be flipped as well as rotated.

H61 Fifteen balls, including three each of five different colors, are arranged in a triangle as shown. How many ways can this be done if rotations of the triangle are allowed?

H62 Consider colorings of the triangular regions in the accompanying figure, allowing both rotations and reflections of the figure.

(a) Decide exactly which transformations are involved and construct the cycle index polynomial.

(b) Find the number of colorings that use two colors.

(c) Find the number of colorings that use three colors, if no two adjacent regions are given the same color.

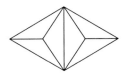

H63 Repeat (a) and (b) of the preceding problem for colorings of the edges (the 11 line segments) in the same figure.

H64 Repeat problem H27(c) for colorings in which no two adjacent squares are given the same color.

H65 Consider colorings of the 24 triangular regions in the accompanying figure, allowing both rotations and reflections.

(a) Construct the cycle index polynomial.

(b) Find the number of colorings that use two colors, with each color occurring on twelve regions.

H66   (a) Construct the cycle index polynomial for coloring the regions in the accompanying figure, allowing both rotations and reflections.

(b) Repeat (a) for coloring the edges in the same figure.

(c) Construct an appropriate cycle index polynomial for colorings that assign colors to both the regions and the edges in this figure. Use the notation X_i for cycles involving regions and Y_i for cycles involving edges.

(d) Find the number of colorings that use three colors for regions and two for edges.

H67   Consider colorings of the nine squares in the accompanying figure that use two colors, allowing rotations but not reflections of the figure, and with the added requirement that opposite corner squares must be the same color.

(a) Construct an appropriate pattern inventory. (*Hint:* The term corresponding to the identity transformation should be $(A + B)^5(A^2 + B^2)^2$.)

(b) Use your answer from (a) to count colorings that use A on five squares, B on four, and assign the same color to opposite corner squares.

Groups of Transformations

Problems H68–70 deal with transformations that rearrange the elements of a finite set. Such transformations form an algebraic structure known as a *group*. For a more general discussion of groups, see any standard text on abstract algebra.

Let S be a finite set and let G be a set of transformations that rearrange the elements of S. Each transformation T in G is a one-to-one function from S onto S. Equivalently, T is a permutation of the elements of S. (Some familiar examples: T can be a rotation of the terms of a string, or a rotation or reflection in two or three dimensions that rearranges parts of a geometrical figure.) We make the further assumption that G is *closed under composition*. This means that whenever two transformations in G (possibly the same transformation, repeated) are applied successively, the result is another transformation in G.

H68 Suppose that G consists of the four reflections of the accompanying figure, along with the identity transformation. (Consider these to be permutations of the four squares.) Show that this set G is *not* closed under composition.

Because S is a finite set, it can be shown that each permutation of S rearranges the elements of S in cycles. For a given transformation T in G, let m denote the least common multiple of all cycle lengths created by T. (For example, if T has cycle code $X_1^5 X_4^2 X_6$, then $m = 12$.) If T is applied m times, the resulting transformation T^m is the identity transformation I.

H69 Be sure you understand why the preceding observation is true.

H70 What is T^{m-1}, in relation to T?

Related to the preceding observations is the fact that if $T^m = I$ for some integer m, then all cycle lengths created by T must be divisors of m. This fact has some very useful consequences, as illustrated in the following problems.

H71 Find the possible cycle lengths created by

(a) a 180° rotation;

(b) a 90° rotation;

(c) a 120° rotation;

(d) any reflection.

H72 Find the number of ways to color the faces of a cube

(a) using five colors;

(b) using three colors, each at least once.

H73 Find the number of ways to color the eight vertices (corner points) of a cube

(a) using two colors;

(b) using two colors, each the same number of times.

H74 Repeat problem H73 for the twelve edges of a cube.

H75 Construct a polynomial $P(X, Y, Z)$ that counts the ways to color the vertices, edges and faces of a cube using X colors for vertices, Y colors for edges, and Z colors for faces. (If necessary, see problem H66.)

H76 Construct the cycle index polynomial for coloring the 24 squares on the surface of the cube shown in the accompanying figure. Find the number of colorings that use two colors.

H77 A cube is divided into 27 smaller cubes as shown in the accompanying figure. Each one of the 26 small cubes excluding the center cube is colored, using two colors. How many ways can this be done?

A *tetrahedron*, shown in the accompanying figure, is a triangular pyramid in which each of the four faces is an equilateral triangle. There are twelve rotations of a tetrahedron: the identity transformation; eight 120° rotations, one in each direction around an axis that runs through a vertex and the center of the opposite face; and three 180° rotations, one around each axis that runs through the midpoints of two opposite edges.

H78 (a) Construct the cycle index polynomial for coloring the faces of a tetrahedron.

(b) Find the number of ways to color the faces of a tetrahedron using k colors, for each $k = 2, 3,$ and 4.

H79 (a) Construct the cycle index polynomial for coloring the edges of a tetrahedron.

(b) Find the number of ways to color the edges of a tetrahedron using three colors, each on two edges.

H80 Three red balls, three blue balls, and four green balls are stacked in a triangular pyramid as shown in the accompanying figure. Find the number of ways to do this if

(a) we allow rotations of the pyramid that keep the same ball on top;

(b) all rotations of the pyramid are allowed. (In this case, consider the balls to be glued together.)

A *dodecahedron*, shown in the accompanying figure, has 12 faces, 30 edges, and 20 vertices. Each face is a regular pentagon. There are 60 rotations of a dodecahedron: the identity transformation; twelve 72° rotations, one in each direction around six axes that run through the centers of two opposite faces; twelve 144° rotations around the same axes; twenty 120° rotations, one in each direction around ten axes that run through two opposite vertices; and fifteen 180° rotations around axes that run through the midpoints of two opposite edges.

H81 (a) Construct the cycle index polynomial for coloring the faces of a dodecahedron.

(b) Find the number of colorings that use two colors, each on six faces.

H82 Repeat problem H81(a) for the edges of a dodecahedron.

H83 Repeat problem H81(a) for the vertices of a dodecahedron.

List of Standard Problems

Standard Problem #1

Find the number of strings of a given length that use elements from a given set.

Standard Problem #2

Find the number of strings of a given length that use elements from a given set if no element appears more than once in any string.

Standard Problem #3

Find the number of combinations of a given length that consist of distinct elements from a given set.

Standard Problem #4

Find the number of bit strings that contain a given number of 0s and 1s, such that there are no two consecutive 1s.

Standard Problem #5

Find the number of combinations of a given length that use elements from a given set, allowing repetition and with no missing elements.

Standard Problem #6
Find the number of combinations of a given length that use elements from a given set, allowing repetition.

Standard Problem #7
Find the number of 0-dominated bit strings that contain a given number of 0s and 1s.

Standard Problem #8
Find the number of distributions of a given set of distinct balls into a given set of distinct boxes. Equivalently, find the number of functions from one set to another.

Standard Problem #9
Find the number of distributions of a given set of identical balls into a given set of distinct boxes.

Standard Problem #10
Find the number of distributions of a given set of identical balls into a given set of distinct boxes, if no boxes are allowed to be empty.

Standard Problem #11
Find the number of distributions of a given set of distinct balls into a given set of distinct boxes, if each box must contain a specified number of balls.

Standard Problem #12
Find the number of rearrangements of a given word.

Standard Problem #13
Find the number of distributions of a set of distinct balls into a set of distinct boxes, if no boxes are allowed to be empty.

Standard Problem #14
Find the number of words of a given length from a given set of letters, if each letter must occur at least once in the word.

Standard Problem #15
For a given set, find the number of partitions of a given type.

Standard Problem #16

For a given set, find the number of partitions that have a specified number of nonempty parts.

Standard Problem #17

Find the number of partitions of a given positive integer that have a specified number of positive parts.

Dependence of Problems

The first entry indicates that problem A24 can be assigned any time after problem A2 has been covered in class; the last entry indicates that problem H83 depends on problem H81.

A24	A2		A44	A7
A25	A3		A45	A22
A26	A7		A46	A23
A27	A10		A47	A23
A28	A8			
A29	A2		B30	B6
A30	A5		B31	B30
A31	A8		B32	B6
A32	A11		B33	B5
A33	A5		B34	B5
A34	A5		B35	B5
A35	A5		B36	B5
A36	A14		B37	B6
A37	A15		B38	B6
A38	A19		B39	B6
A39	A17		B40	B12
A40	A18		B41	B13
A41	A18		B42	B14
A42	A18		B43	B10
A43	A42		B44	B5

B45	B44		C49	C30
B46	B5		C50	C34
B47	B5		C51	C30
B48	B15		C52	C30
B49	B17		C53	C33
B50	B17		C54	C33
B51	B23		C55	C34
B52	B23		C56	C55 [lozenge]
B53	B23		C57	C56
B54	B23		C58	C57
B55	B16		C59	C34
B56	B55		C60	C35
B57	B56		C61	C35
B58	B28		C62	C34
B59	B28		C63	C37
B60	B28		C64	C39
B61	B28		C65	C64
B62	B58			
B63	B58		D36	D4
B64	B6		D37	D11
B65	B64		D38	D15
B66	B65		D39	D15
B67	B6		D40	D16
B68	B23		D41	D16
B69	B68		D42	D17
B70	B6		D43	D17
B71	B70		D44	D27
B72	B12		D45	D33
B73	B72		D46	D22
			D47	D24
C40	C12		D48	D35
C41	C12		D49	D21
C42	C14		D50	D49
C43	C17		D51	D50
C44	C24		D52	D50
C45	C24		D53	D16
C46	C26		D54	D53
C47	C22		D55	D54
C48	C31		D56	D53

D57	D56		F33	F32
D58	D57		F34	F31
D59	D58		F35	F15
D60	D59		F36	F16
D61	D60		F37	F17
D62	D61		F38	F19
			F39	F16
E31	E8		F40	F16
E32	E8		F41	F19
E33	E10		F42	F16
E34	E8		F43	F10
E35	E34		F44	F5
E36	E35		F45	F44
E37	E11		F46	F45
E38	E37		F47	F17
E39	E38		F48	F22
E40	E11		F49	F5
E41	E11		F50	F49
E42	E7		F51	F5
E43	E9		F52	F51
E44	E14			
E45	E14		G24	G4
E46	E14		G25	G24
E47	E7		G26	G5
E48	E12		G27	G5
E49	E20		G28	G8
E50	E20		G29	G8
E51	E50		G30	G13
E52	E28		G31	G15
			G32	G15
F24	F5		G33	G15
F25	F5		G34	G21
F26	F25		G35	G21
F27	F15		G36	G23
F28	F15		G37	G21
F29	F28		G38	G21
F30	F15		G39	G38
F31	F30		G40	G39
F32	F31			

H50	H2
H51	H13
H52	H26
H53	H52
H54	H30
H55	H54
H56	H27
H57	H26
H58	H30
H59	H43
H60	H42
H61	H42
H62	H42
H63	H62
H64	H27
H65	H42
H66	H42
H67	H42
H68	H40
H69	H68
H70	H69
H71	H70
H72	H48
H73	H49
H74	H73
H75	H74
H76	H48
H77	H48
H78	H48
H79	H78
H80	H78
H81	H48
H82	H81
H83	H81

Answers to Selected Problems

A6 $5!/2 = 60$

A8 $26 \cdot 25^4$

A11   There are $2^5 = 32$ words that contain no A, 32 words that contain no B, and 32 words that contain no C. Among these, three ($AAAAA$, $BBBBB$, and $CCCCC$) are counted twice, so the number of words that have at least one missing letter is 93. Therefore the answer is $3^5 - 93 = 150$.

A13 $(10 \cdot 9 \cdot 8)/6 = 120$

A18   (a) $4!/(52 \cdot 51 \cdot 50 \cdot 49)$

       (b) $(52 \cdot 39 \cdot 26 \cdot 13)/(52 \cdot 51 \cdot 50 \cdot 49)$

A22 $3^4 = 81$. These are not derangements because repetition is allowed.

A23 $D_4 = 9$. These are derangements of 2314.

B2   (a) 10

       (b) 35

B6

	1	6	15	20	15	6	1	
1	7	21	35	35	21	7	1	
1	8	28	56	70	56	28	8	1
1	9	36	84	126	126	84	36	9	1

B10 $\binom{7}{3} = 35$

B13 (a) $-\binom{15}{8}$

    (b) $\binom{15}{4}$

B16 $\binom{m+1}{n}$

B17 $\binom{8}{3} = 56$. There is a one-to-one correspondence between combinations of three nonconsecutive digits and the sequences counted in problem B15: Replace each selected digit by an A and each nonselected digit by a B.

B20 $\binom{39}{9}$ Start with one of each digit and apply Standard Problem #5 with $k = 40$.

B24 $\binom{100}{10}_R = \binom{109}{10}$ Each sequence corresponds to a combination allowing repetition.

B27 (a) $AABBBA$

    (b) $BABABA$

    (c) $BAABAA$

    (d) $BABAAA$

C1 Each ball determines a letter according to which box it goes into. This distribution corresponds to the word $BAACB$.

C6 $8 \cdot 7 \cdot 6 \cdot 5 \cdot 4$

C8 There are $2^5 = 32$ distributions in which box A is empty, 32 in which B is empty, and 32 in which C is empty. Among these, three distributions (in which all balls go into one box) are counted twice, so the number of distributions in which at least one box is empty is 93. Therefore the answer is $3^5 - 93 = 150$.

C12 $\binom{3}{5}_R^2 = \binom{7}{2}^2 = 441$

C15 $\binom{35}{3}$. Let $m = 36$ in Standard Problem #10.

C18 2520

C24 (a) 20

    (b) 12

    (c) 40

    (d) 12

C26 (a) $\binom{7}{2,3,2}$

    (b) $-\binom{7}{1,3,3}$

C32 $\binom{5}{3}T(5,3) = 1500$

C34   1 62 540 1560 1800 720
        1 126 1806 8400 16800 15120 5040

C37 $n(n + 1) \cdots (n + m - 1)$

D2   Each partition corresponds to six distributions.

D10   1
        1 1
       1 3 1
      1 7 6 1
     1 15 25 10 1

D11   1 31 90 65 15 1
       1 63 301 350 140 21 1
      1 127 966 1701 1050 266 28 1

D15   (a) 52
      (b) 36
      (c) 41

D17 $3 \cdot S(5, 3) = 75$

D21   1
        1 1
       1 1 1
      1 2 1 1
     1 2 2 1 1
    1 3 3 2 1 1

D27   1 3 4 3 2 1 1
       1 4 5 5 3 2 1 1
      1 4 7 6 5 3 2 1 1
     1 5 8 9 7 5 3 2 1 1

D30 $x = m + n$; partitions of m having at most n positive parts correspond to partitions of m having exactly n nonnegative parts, which in turn correspond to partitions of $m + n$ having exactly n positive parts.

D34   (a) $1 + \cdots + (n - 1) + (n - 1) + \cdots + 1 = n(n - 1)$
      (b) $x = m - \binom{n}{2}$; subtract 1 from the second term, 2 from the third, etc.

D35   15

E1   Elements in exactly one of the three sets get counted once by the given expression. Elements in exactly two sets get counted twice by the first three terms and then are subtracted once in the next group of terms, for a net contribution of 1. Elements in all three sets get counted three times, subtracted three times, and finally counted once more by the final term.

E4 (a) Let $A = \{\text{rearrangements with 1 in position 1}\}$
$\qquad B = \{\text{rearrangements with 2 in position 2}\}$
$\qquad C = \{\text{rearrangements with 3 in position 3}\}$
$\qquad D = \{\text{rearrangements with 4 in position 4}\}$
Then $|A \cup B \cup C \cup D| = 4(3!) - 6(2!) + 4 - 1 = 15$

(b) $4! - 15 = 9$

E5 (a) 44

(b) 16

E8 $3^m - 3(2^m) + 3$

E10 $441 - 108 + 3 = 336$

E12 $2520 - 2520 + 1080 - 240 + 24 = 864$

E14 $120 - 96 + 36 - 8 + 1 = 53$

E15 $100000 - 3000 + 210 - 10 = 97200$

E16 (a) $5\binom{5}{2}_R = 75$

(b) $\binom{5}{5}_R - 75 = 126 - 75 = 51$

E17 $462 - 336 + 15 = 141$

E25 $S_1 - 2S_2 + 3S_3 - \cdots \pm nS_n$

E26 (a) $972 - 384 + 12 = 600$

(b) $4 \cdot T(5, 3) = 600$

E29 $N_3 = S_3 - 4S_4 + 10S_5 - 20S_6 + \cdots \pm \binom{n}{3}S_n$

F3 $1, 2, 3, 5$

F5 81

F6 The number of As and Bs is not specified here.

F11 144

F13 (a) Words starting with A must have B next. Words starting with B continue with A or BA.

(b) $a_n = b_{n-1} = a_{n-2} + a_{n-3}$
$\qquad b_n = a_{n-1} + a_{n-2} = b_{n-2} + b_{n-3}$
$\qquad w_n = a_n + b_n = a_{n-2} + a_{n-3} + b_{n-2} + b_{n-3} = w_{n-2} + w_{n-3}$

(c) 49

F15

n	1	2	3	4	5
a_n	1	4	20	95	456
b_n	1	5	23	111	532
c_n	1	5	24	115	551
w_n	5	24	115	551	2641

F18 $A = (1 + \sqrt{5})/2\sqrt{5}, B = (\sqrt{5} - 1)/2\sqrt{5}, F_{10} = 89, F_{25} = 121393$

F21   If $x_1 \ldots x_n$ is a derangement of $1 \ldots n$, let $x_1 = k$. If $x_k = 1$, then $x_2 \ldots x_{k-1} x_{k+1} \ldots x_n$ is a derangement of $2, \ldots, k - 1, k + 1, \ldots n$. If $x_k \neq 1$, then $x_2 \ldots x_n$ is a derangement of $2, \ldots, k - 1, 1, k + 1, \ldots n$.

G3 $(1 + X)^4 = 1 + 4X + 6X^2 + 4X^3 + X^4$

G5 $(R + R^2 + R^3)(1 + B^2 + B^4)(1 + G + G^2) \longrightarrow X(1 + X + X^2)^2(1 + X^2 + X^4)$
The coefficient of X^5 is 5.

G9   12, 10

G14 $(1 + X)(1 + X^2)(1 + X^3)(1 + X^4)(1 + X^5)(1 + X^6)$

G17   19

G18   7

G21   1, 3 ,8, 19, 38, 60, 60

G22   65

G23   85

H2 $\{AAABBB, AABBBA, ABBBAA, BBBAAA, BBAAAB, BAAABB\}$,
$\{AABAAB, ABAABA, BAABAA\}$, $\{ABABAB, BABABA\}$
All orbit lengths are divisors of 6.

H3   (a) 5

     (b) Orbit lengths are divisors of n. Look for the smallest repeating block in the string.

H6   Six ways, one for each orbit among the four-letter strings that use the letters A and B.

H7   11

H8   8

H9   6; each orbit of length n contributes n terms to the sum, each of which is $1/n$.

H10   The terms are rearranged. The same is true for each rotation.

H13 The $R_i(X_1 \ldots X_n)$ run through different rotations of $X_1 \ldots X_n$ until the original string is repeated and the process begins again. Each time the original string repeats, another copy of the orbit begins.

H14 $4 + 1 + 1 + 1 + 1 + 2 + 2 + 1 + 1 + 1 + 1 + 1 + 1 + 1 + 1 + 4 = 24$

H16 Each sum counts ordered pairs (R, S) consisting of a rotation R and a string S that is unchanged by R.

H17 $(32 + 2 + 2 + 2 + 2)/5 = 8$

H18 11

H19 24

H22 51

H23 $(64 + 2 + 4 + 8 + 4 + 2)/6 = 14$

H24 Each d-letter initial segment determines an invariant string.

H25 $r_{10} = r_5 = 3^5 = 243$

H26 $(T(5, 3) + 0 + 0 + 0 + 0)/5 = 30$

H27 (a) 3^4, 3, 3^2, 3 for rotations; 3^2 for reflections through horizontal and vertical axes; and 3^3 for reflections through diagonals.

   (b) 24

   (c) 21

H29 39

H30 13

H31 (a) Each rotation determines groups of squares that must be given the same color in any invariant coloring. The exponent is the number of these groups.

   (b) $(2^7 + 2^3 + 2^5 + 2^3)/4 = 140$

H35 Each term in the expression comes from a choice of either A or B from the first factor and either A^2 or B^2 from each of four more factors. These choices correspond to selecting either A or B for the center square and either A or B for each of the four pairs of opposite squares.

H37 $X_1^9, X_1 X_2^4, X_1 X_4^2$

H40 (a) $(X_1^4 + 2X_4 + X_2^2)/4$

   (b) $((A + B)^4 + 2(A^4 + B^4) + (A^2 + B^2)^2)/4$

   (c) There is one coloring that uses A on exactly k squares, for each $k = 0, 1, 3$, and 4. For $k = 2$ there are two colorings. The expansion of the pattern inventory is

$$A^4 + A^3 B + 2A^2 B^2 + AB^3 + B^4$$

H41 $(X_1^4 + 2X_4 + 3X_2^2 + 2X_1^2X_2)/8$

H43 (a) $(X_1^6 + 2X_6 + 2X_3^2 + 4X_2^3 + 3X_1^2X_2)/12 = 13$ when all $X_i = 2$

(b) The coefficient of B^3W^3 in the pattern inventory is 3. The types can be represented by

$$BBBWWW, \quad BBWBWW, \text{ and } \quad BWBWBW.$$

H47 $(X_1^6 + 6X_1^2X_4 + 3X_1^2X_2^2 + 8X_3^2 + 6X_2^3)/24$

H48 (a) 10

(b) 57

Index

binary system, 3
binomial expansion, 17
birthday problem, 10, 11
bit, 3
bit string, 9, 18
Burnside, W., 111

cards, 10, 25, 44, 45, 66, 81
chessboard, 9
circular arrangement, 9, 26, 58, 59, 102,
    104, 112
coin flipping, 10, 24
coin problem, 89
combination, 13, 14, 23
combination allowing repetition,
    18–21, 23
combination lock, 9
combination number, 14–16, 23
combination with limited repetition, 67,
    87
consecutive repetition, 6, 18, 75
consistently dominated sequence, 21
counting paths, 22, 23

counting regions, 84
cycle code, 107, 108, 111
cycle index polynomial, 107, 108, 111
cycle polynomial, 107, 108

derangement, 7, 8, 11
derangement number, 8, 64, 71, 74,
    79–81
distribution, 31, 32, 35, 42
distribution into identical boxes, 52
distribution number, 34, 42
distribution of identical objects, 32–34,
    42
distribution of type $(m_1, \ldots, m_n)$, 35, 48
divisibility, 11, 28, 49, 64, 71, 72

election problem, 27
exponential generating function,
    90–93

Fibonacci sequence, 74, 79, 81, 83
flagpole problem with distinct poles,
    41, 43